[Ceramic Insulator]

ELECTRICAL PYLON

Many of the things you use every day run on electricity. Electricity often comes from the outlet in the wall where you plug in the cord. But from where does that electricity come?

You may have noticed tall towers near your home. These tall towers are called electrical pylons. The pylons have power lines that bring electricity all the way from a power station to the electrical wires that run through your home. The next time your alarm goes off, think about how far that electricity has traveled!

The part of the tower that connects to the electrical wires is made of ceramic, which keeps people who work on the tower safe from the many dangers of electricity.

NATIONAL GEOGRAPHIC
SCIENCE

PHYSICAL SCIENCE

NATIONAL GEOGRAPHIC

School Publishing

PROGRAM AUTHORS

Malcolm B. Butler, Ph.D.

Judith S. Lederman, Ph.D.

Randy Bell, Ph.D.

Kathy Cabe Trundle, Ph.D.

David W. Moore, Ph.D.

Program Authors

MALCOLM B. BUTLER, PH.D.

Associate Professor of Science Education,
University of South Florida, St. Petersburg, Florida
SCIENCE

JUDITH SWEENEY LEDERMAN, PH.D.

Director of Teacher Education,
Associate Professor of Science Education,
Department of Mathematics and Science Education,
Illinois Institute of Technology, Chicago, Illinois
SCIENCE

RANDY BELL, PH.D.

Associate Professor of Science Education,
University of Virginia, Charlottesville, Virginia
SCIENCE

KATHY CABE TRUNDLE, PH.D.

Associate Professor of Early Childhood Science
Education, The School of Teaching and Learning,
The Ohio State University, Columbus, Ohio
SCIENCE

DAVID W. MOORE, PH.D.

Professor of Education,
College of Teacher Education and Leadership,
Arizona State University, Tempe, Arizona
LITERACY

Program Reviewers

Amani Abuhabsah
Teacher
Dawes Elementary
Chicago, IL

Maria Aida Alanis, Ph.D.
Elementary Science
Instructional Coordinator
Austin Independent
School District
Austin, TX

Jamillah Bakr
Science Mentor Teacher
Cambridge Public Schools
Cambridge, MA

Gwendolyn Battle-Lavert
Assistant Professor of Education
Indiana Wesleyan University
Marion, IN

Carmen Beadles
Retired Science Instructional
Coach
Dallas Independent School
District
Dallas, TX

Andrea Blake-Garrett, Ed.D.
Science Educational Consultant
Newark, NJ

Lori Bowen
Science Specialist
Fayette County Schools
Lexington, KY

Pamela Breitberg
Lead Science Teacher
Zapata Academy
Chicago, IL

Program Reviewers continued
on page iv.

Acknowledgments

Grateful acknowledgment is given to the authors, artists, photographers, museums, publishers, and agents for permission to reprint copyrighted material. Every effort has been made to secure the appropriate permission. If any omissions have been made or if corrections are required, please contact the Publisher.

Illustrator Credits
All illustrations by Precision Graphics. All maps by Mapping Specialists.

Photographic Credits
Front Cover Chris Knapton/Photo Researchers, Inc.

Credits continue on page EM16.

The National Geographic Society

John M. Fahey, Jr.,
President & Chief Executive Officer

Gilbert M. Grosvenor,
Chairman of the Board

Copyright © 2011 The Hampton-Brown Company, Inc., a wholly owned subsidiary of the National Geographic Society, publishing under the imprints National Geographic School Publishing and Hampton-Brown.

National Geographic School Publishing
Hampton-Brown
www.myNGconnect.com

Printed in the USA.
RR Donnelley
Jefferson City, MO

ISBN: 978-0-7362-7724-2

12 13 14 15 16 17 18 19 20

6 7 8 9 10

Carol Brueggeman
K–5 Science/Math Resource
Teacher
District 11
Colorado Springs, CO

Miranda Carpenter
Teacher, MS Academy Leader
Imagine School
Bradenton, FL

Samuel Carpenter
Teacher
Coonley Elementary
Chicago, IL

Diane E. Comstock
Science Resource Teacher
Cheyenne Mountain School
District
Colorado Springs, CO

Kelly Culbert
K–5 Science Lab Teacher
Princeton Elementary
Orange County, FL

Karri Dawes
K–5 Science Instructional
Support Teacher
Garland Independent
School District
Garland, TX

Richard Day
Science Curriculum Specialist
Union Public Schools
Tulsa, OK

Michele DeMuro
Teacher/Educational Consultant
Monroe, NY

Richard Ellenburg
Science Lab Teacher
Camelot Elementary
Orlando, FL

Beth Faulkner
Brevard Public Schools
Elementary Training Cadre,
Science Point of Contact, Teacher,
NBCT
Apollo Elementary
Titusville, FL

Kim Feltre
Science Supervisor
Hillsborough School District
Newark, NJ

Judy Fisher
Elementary Curriculum
Coordinator
Virginia Beach Schools
Virginia Beach, VA

Anne Z. Fleming
Teacher
Coonley Elementary
Chicago, IL

Becky Gill, Ed.D.
Principal/Elementary Science
Coordinator
Hough Street Elementary
Barrington, IL

Rebecca Gorinac
Elementary Curriculum Director
Port Huron Area Schools
Port Huron, MI

Anne Grall Reichel Ed. D.
Educational Leadership/
Curriculum and Instruction
Consultant
Barrington, IL

Mary Haskins, Ph.D.
Professor of Biology
Rockhurst University
Kansas City, MO

Arlene Hayman
Teacher
Paradise Public School District
Las Vegas, NV

DeLene Hoffner
Science Specialist, Science
Methods Professor,
Regis University
Academy 20 School District
Colorado Springs, CO

Cindy Holman
District Science Resource Teacher
Jefferson County Public Schools
Louisville, KY

Sarah E. Jesse
Instructional Specialist for
Hands-on Science
Rutherford County Schools
Murfreesboro, TN

Dianne Johnson
Science Curriculum Specialist
Buffalo City School District
Buffalo, NY

Kathleen Jordan
Teacher
Wolf Lake Elementary
Orlando, FL

Renee Kumiega
Teacher
Frontier Central School District
Hamburg, NY

Edel Maeder
K–12 Science Curriculum
Coordinator
Greece Central School District
North Greece, NY

Trish Meegan
Lead Teacher
Coonley Elementary
Chicago, IL

Donna Melpolder
Science Resource Teacher
Chatham County Schools
Chatham, NC

Melissa Mishovsky
Science Lab Teacher
Palmetto Elementary
Orlando, FL

Nancy Moore
Educational Consultant
Port Stanley, Ontario, Canada

Melissa Ray
Teacher
Tyler Run Elementary
Powell, OH

Shelley Reinacher
Science Coach
Auburndale Central Elementary
Auburndale, FL

Kevin J. Richard
Science Education Consultant,
Office of School Improvement
Michigan Department of
Education
Lansing, MI

Cathe Ritz
Teacher
Louis Agassiz Elementary
Cleveland, OH

Rose Sedely
Science Teacher
Eustis Heights Elementary
Eustis, FL

Robert Sotak, Ed.D.
Science Program Director,
Curriculum and Instruction
Everett Public Schools
Everett, WA

Karen Steele
Teacher
Salt Lake City School District
Salt Lake City, UT

Deborah S. Teuscher
Science Coach and
Planetarium Director
Metropolitan School District
of Pike Township
Indianapolis, IN

Michelle Thrift
Science Instructor
Durrance Elementary
Orlando, FL

Cathy Trent
Teacher
Ft. Myers Beach Elementary
Ft. Myers Beach, FL

Jennifer Turner
Teacher
PS 146
New York, NY

Flavia Valente
Teacher
Oak Hammock Elementary
Port St. Lucie, FL

Deborah Vannatter
District Coach, Science Specialist
Evansville Vanderburgh School
Corporation
Evansville, IN

Katherine White
Science Coordinator
Milton Hershey School
Hershey, PA

Sandy Yellenberg
Science Coordinator
Santa Clara County Office
of Education
Santa Clara, CA

Hillary Zeune de Soto
Science Strategist
Lunt Elementary
Las Vegas, NV

PHYSICAL SCIENCE

CONTENTS

TECHTREK
myNGconnect.com

Student eEdition | Vocabulary Games | Digital Library | Enrichment Activities

CHAPTER
3

TECHTREK
myNGconnect.com

Student
eEdition

Vocabulary
Games

Digital
Library

Enrichment
Activities

CHAPTER

5

PHYSICAL SCIENCE

What Is Physical Science?

Physical science is the study of the physical world around you. This type of science investigates the properties of different objects, as well as how those objects interact with each other. Physical science includes the study of matter, motion and forces, and many kinds of energy, including light and electricity. People who study how all of these things work together are called physical scientists.

You will learn about these aspects of physical science in this unit:

HOW CAN YOU DESCRIBE AND MEASURE MATTER?

Matter is anything that has mass and takes up space. Physical scientists study all of the different properties of matter. These include size, shape, color, texture, and hardness, as well as mass and volume.

WHAT ARE STATES OF MATTER?

Matter can be found in many different states. Each of these states has its own special properties. Physical scientists study these different states of matter. They also study how temperature changes can cause matter to change from one state of matter to another.

HOW DOES FORCE CHANGE MOTION?

Physical scientists study how objects move. They also study the forces that act on objects, and how those objects respond to different kinds of forces. Physical scientists also observe and measure the changes that forces cause.

WHAT IS ENERGY?

Physical scientists study energy in all of its forms. They also learn about how energy can cause changes in the physical world. The different kinds of energy include light, sound, electrical, and mechanical.

WHAT IS LIGHT?

Light is a kind of energy that you can see. Physical scientists study the properties of light. They also learn about how light can change direction and even cause objects to heat up.

MEET A SCIENTIST

Constance Adams: Space Architect

Constance Adams is a space architect with NASA and a National Geographic Emerging Explorer. One of her first projects with NASA was TransHab, designed to be a transit habitat for the first human mission to Mars.

Constance and her team of experts bring together innovations from diverse disciplines such as architecture, engineering, industrial design, and sociology with the hopes of solving complex design issues. "When you have a brand-new problem, you need as many tools as you can get. Who knows—an approach from a very different field might give you the insight you need. For example, I'm working to forge communication between advanced engineering and consumer-product design to bring more user-centered designs to aerospace," says Constance.

As a space architect, Constance works with space agencies, including NASA, to design and build components for spacecraft and the International Space Station.

CHAPTER

1

HOW CAN YOU

DESCRIBE AND

MEASURE

MATTER?

Take a look around you. What do you see? Matter! Everything around you is made of matter. These hot air balloons are made of matter. You can use your senses to describe them. The balloons are colorful and big. You can measure matter, too.

TECHTREK
myNGconnect.com

Pick a balloon and describe it. How could you measure the balloon?

6   7

After reading Chapter 1, you will be able to:

- Recognize that all objects and substances in the world are made of matter.
  PROPERTIES OF MATTER

- Recognize that matter takes up space and has mass. **PROPERTIES OF MATTER**

- Identify the properties of objects and substances, such as size, shape, color, texture, and hardness. **PROPERTIES OF MATTER**

- Show that the property of color may be dependent on the surroundings in which the object exists. **PROPERTIES OF MATTER**

- Recognize that materials may be composed of parts too small to be seen without magnification. **PROPERTIES OF MATTER**

- Explain how to measure and compare the mass and volume of solids and liquids.
  MEASURING MASS, MEASURING VOLUME

- Science in a Snap! Show that the property of color may be dependent on the surroundings in which the object exists. **PROPERTIES OF MATTER**

HOW CAN YOU DESCRIBE MEA

Take a look around you. What do you see? Matter! Everything around you is made of matter. These hot air balloons are made of matter. You can use your senses to describe them. The balloons are colorful and big. You can measure matter, too.

Student
eEdition

Vocabulary
Games

Digital
Library

Enrichment
Activities

AND SURE MATTER?

Pick a balloon and describe it. How could you measure the balloon?

SCIENCE VOCABULARY

matter (MA-ter)

Matter is anything that has mass and takes up space. (p. 10)

The airplane and buildings are made of matter.

texture (TEKS-chur)

Texture describes the surface of any area made up of matter. (p. 14)

The metal armrest has a smooth texture.

my

Science Vocabulary

mass
(MAS)

matter
(MA-ter)

texture
(TEKS-chur)

volume
(VOL-yum)

TECHTREK
myNGconnect.com

Vocabulary
Games

mass (MAS)

Mass is the amount of matter in an object. (p. 16)

> The items that hang lower have a greater mass.

volume (VOL-yum)

Volume is the amount of space matter takes up. (p. 18)

> The city checks the volume in the tank to be sure there is enough fresh water.

Color and Shape   When you leave an airplane, you have to find your suitcase. How do you find it? You probably observe the properties of color and shape. These suitcases are different shapes and colors. Which do you think would be easiest to find?

TECHTREK
myNGconnect.com

Digital Library

Color is a property you observe with your sense of sight.

Science in a Snap! Changing Colors

Shine a flashlight onto a white piece of paper. Then cover the flashlight with a piece of colored cellophane.

Shine the flashlight on a white piece of paper.

What happens to the color of the light?

Shape is important! The round wheels on these suitcases help the family move quickly through the airport. Square wheels wouldn't work.

Texture and Hardness Airplane passengers want comfortable seats. That's why the chairs are covered with smooth fabric like cotton instead of scratchy fabric like wool. *Smooth* and *scratchy* are words that describe **texture**. Texture is a property of an object's surface you feel by touching.

If you look at cotton and wool from a distance, they look very similar. If you look at them up close with a magnifying glass, though, the wool looks bumpier than the cotton.

bumpy wool

smooth cotton

What textures do you think you could describe on this airplane?

If you pushed your finger against an airplane seat, your finger would make a dent. If you pushed your finger against the outside of an airplane, your finger could not press into the surface. The outside of an airplane is harder than an airplane seat. Hardness is another property that describes matter.

The soft seat gives the passenger a comfortable ride.

soft

hard

Before You Move On

1. Name properties that you can use to describe matter.
2. Choose an object in the classroom and describe its properties.
3. **Compare** How is texture different from hardness? How can you observe both properties?

Measuring Mass

You can compare sizes of objects by measuring their **mass**. Mass is the amount of matter in an object. Mass can be measured in grams or ounces.

A balance can help you compare the masses of different solids. The objects with the greater mass make one tray hang lower. However, a balance cannot tell you the weight of the objects. This is because mass and weight are not the same thing. An object weighs more when there is more gravity. Earth has more gravity than the moon, so you would weigh more on Earth. You would have the same amount of matter, though, in both places. Your mass on the moon would be the same as your mass on Earth.

TECHTREK
myNGconnect.com

Enrichment Activities

Which has a greater mass—the stones or the feathers? Explain how you know.

How do you measure the mass of a liquid? First, measure the mass of an empty container. Add the liquid to the container. Then, measure the mass of the container and the liquid. Subtract the first measurement from the second measurement. The remainder is the mass of the liquid.

It is important to measure the mass of the container before adding the liquid.

Measure the mass of the container and liquid. What should you do next?

Before You Move On

1. Identify the units for measuring the mass of an object.
2. Describe how to measure the mass of a liquid.
3. **Compare** How does your mass on Earth compare to your mass on the moon? How does your weight on Earth compare with your weight on the moon?

Measuring Volume

Which holds more water, the water bottle or the water tank? The tank holds more water, but how could you figure out exactly how much more? You could compare **volume** . Volume is a property of matter. The amount of space that matter takes up is its volume. Liquid volume is measured in liters or gallons.

TECHTREK
myNGconnect.com

Digital Library

The volume of the bottle is one liter. The volume of the tank is many thousands of liters.

Look at the two containers below. Do you think they are holding the same volume of liquid? The volume of liquid in one container looks greater than the volume in the other container. Both containers hold the same amount of liquid, though. The shape of a container can fool our eyes into thinking there is more liquid in one container than in another.

You can use different sized beakers to measure the volumes of liquids.

You can use math to figure out the volume of a solid. First, measure the solid's length (l). Length means how long something is. Measure its width (w). Width means how wide something is. Measure its height (h). Height means how tall something is. Then, multiply length by width by height. The result is the volume of the solid.

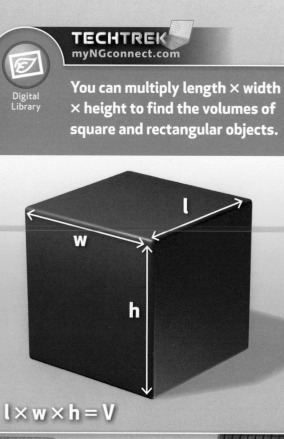

TECHTREK
myNGconnect.com

Digital Library

You can multiply length × width × height to find the volumes of square and rectangular objects.

l × w × h = V

If you wanted to know the volume of one of these buildings, you would need to measure its height, length, and width.

If you want to find the volume of a solid block, you can measure the length, width, and height. What if you want to find the volume of an object with an odd shape, such as a shell? Put water in a beaker. Measure its volume. Then, put the object in the beaker. Measure the volume again to see how far the water rose. The difference in the water's volume is the volume of the object.

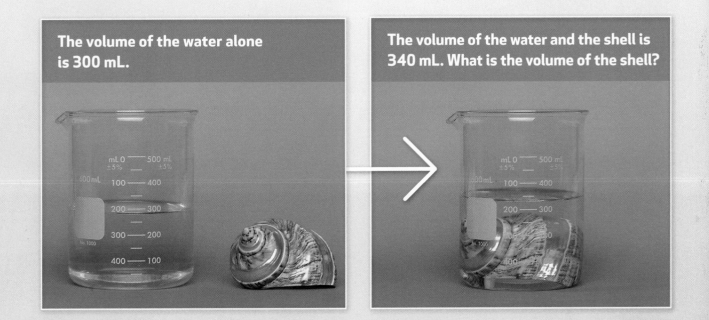

The volume of the water alone is 300 mL.

The volume of the water and the shell is 340 mL. What is the volume of the shell?

Before You Move On

1. Define volume.
2. Summarize how to measure the volume of a liquid.
3. **Describe** Suppose you have a solid marble. You want to measure its volume. Describe what you would do.

GIANTS OF THE PUMPKIN PATCH

You can describe and measure things that people make and things that you find in nature. How would you describe a pumpkin that weighs almost as much as a small car? You might describe it as amazing or gigantic. Scientists, however, would focus on its shape, color, texture, and mass. The mass of one of the largest pumpkins on record is 768 kilograms. That's a whopping 1,689 pounds!

How would you describe these pumpkins using the properties of size, color, and texture?

Small pumpkins often have round shapes. Giant pumpkins may have lumpy shapes. Giant pumpkins grow in different colors, from orange to white. They are hard, and they usually have smooth textures, at least on the outside. The texture of the material on the inside is stringy.

You might think that a pumpkin that has the same mass as an adult grizzly bear would make a lot of pumpkin pies. The pies would not taste good, though. People who grow giant pumpkins don't grow them for food. Instead, they enter the pumpkins in contests. They just want to grow the largest pumpkins they can!

This pumpkin did not win a size prize because of a tiny crack in the bottom.

Matter is anything that has mass and takes up space. You can describe matter by telling about its properties, such as shape, color, texture, and hardness. You can use tools to measure the mass and volume of matter.

Big Idea Matter can be described and measured.

SOME PROPERTIES OF MATTER

Color	Size	Shape	Texture

Vocabulary Review

Match the following terms with the correct definition.

A. matter   **1.** The amount of matter in an object

B. mass   **2.** Describes the surface of any area made up of matter

C. volume

D. texture   **3.** Anything that has mass and takes up space

            4. The amount of space matter takes up

Big Idea Review

1. **Define** Tell what a property is. How can properties help you tell more about objects?

2. **Explain** Suppose you wanted to measure the volume of this book. How would you do it?

3. **Summarize** A friend asks why it's important to describe the properties of objects. How will you respond?

4. **Compare and Contrast** How are the texture of an apple and an orange different? What properties do the two objects have in common?

5. **Infer** You touch an object, and your finger makes a dent in it. What property of matter are you testing? How would you describe the object?

6. **Analyze** Suppose you had a pile of rocks and you wanted to sort them. What properties might the rocks have in common? What properties might be different?

Write About Matter

Explain Observe the object below. Describe its properties. Explain how you would measure its volume and mass.

PHYSICAL SCIENCE EXPERT: PHYSICAL CHEMIST

Rod Ruoff

Rod Ruoff is a physical chemist. He works at the University of Texas at Austin. He also started a company where he works with his team to research new ways of making energy. The team uses materials with interesting properties.

NG Science: What do you currently study?

Rod Ruoff: My team and I study properties of materials. We think about how these materials can make society better.

NG Science: When did you first know you wanted to be a physical chemist?

Rod Ruoff: Actually, I wanted to be either a professional hockey player or soccer player! In college, I took a physical chemistry course. I became fascinated with looking at different materials and thinking about new ways to use them.

NG Science: What type of research have you done?

Rod Ruoff: At first, I studied tiny pieces of matter to learn about how they stick together. Now I mostly research materials made of carbon. Diamonds are made of carbon, and so is graphite. Graphite is the same material as the "lead" in your pencil.

Individual layers of graphite are called "graphene" and we even make and study such atom-thick layers! I want to make materials that can help our environment.

NG Science: Why is your research important?

Rod Ruoff: We think of new ideas that no one has ever thought of before. Our ideas can help others. We tackle hard problems.

NG Science: What is your favorite thing about being a physical chemist?

Rod Ruoff: There are many favorite things for me as a scientist. I think solving important problems, either alone or with a team, brings me the most joy.

TECHTREK
myNGconnect.com

Rod Ruoff and a team member study the shape of matter.

Digital Library

BECOME AN EXPERT

The Properties of Graphene

Graphene is an unusual type of **matter**. Graphene has something in common with diamonds and with the graphite that makes up your pencil lead. All these materials are a form of matter called carbon. Each form of carbon has different properties you can observe.

TECHTREK
myNGconnect.com

Digital
Library

FORMS OF CARBON

Graphene, graphite, and diamonds are different forms of carbon.

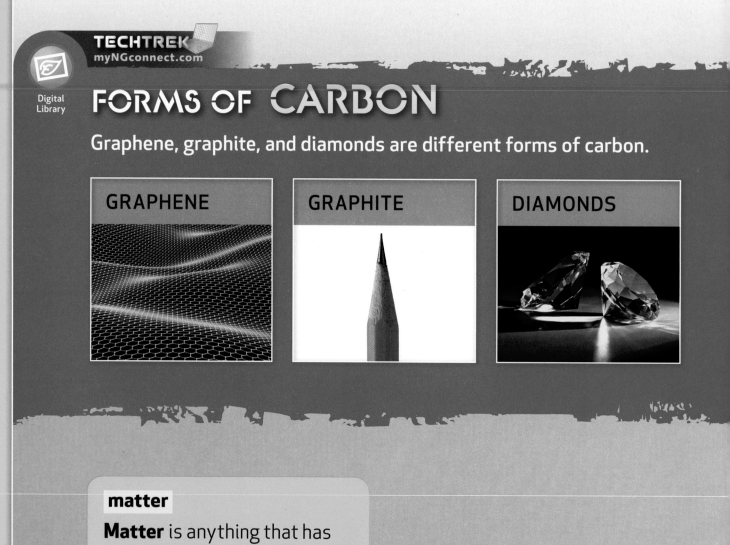

GRAPHENE

GRAPHITE

DIAMONDS

matter
Matter is anything that has mass and takes up space.

TECHTREK
myNGconnect.com

Student
eEdition

Digital
Library

Color Think about the different forms of carbon. What color are they? Natural carbon is very dark. So is the graphite in pencil lead. When you write, the dark graphite wears off the pencil and onto your paper. Diamonds have no color—they are clear. Like graphite, graphene is dark.

Graphite

Size Unlike diamonds and graphite, graphene is so thin, it's almost invisible! A sheet of graphene is less than one nanometer thick. Look at the photo of the ruler. Do you see how small 1 millimeter is? It takes 1 million nanometers to make 1 millimeter!

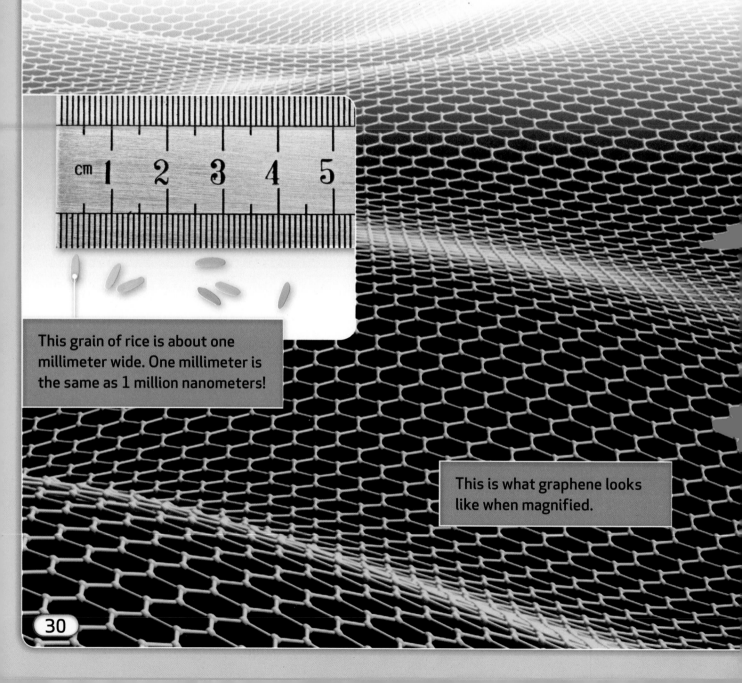

This grain of rice is about one millimeter wide. One millimeter is the same as 1 million nanometers!

This is what graphene looks like when magnified.

Even with its tiny size, graphene is incredibly strong. It's the strongest material ever tested. Think of a sheet of paper. Paper is not very strong. You can rip a sheet in half. Graphene is much thinner than paper. You might think you could rip a sheet of graphene in half, too. Graphene, though, is almost 200 times as strong as steel!

The steel beam is much stronger than this sheet of paper.

Graphene is 200 times as strong as steel.

Hardness The graphite form of carbon is soft. The diamond form of carbon, though, is one of the hardest materials on Earth. Graphene is a hard material, too. Adding graphene to materials can make them stiff. If you hold a sheet of paper on the side by its edges, it will bend toward the floor. A sheet of graphene is much thinner than a sheet of paper. If you could hold a sheet of graphene, it would be stiff. Researchers are looking for ways to make materials stronger and stiffer by adding graphene to them.

Graphene is much thinner than this paper, but a sheet of graphene would not bend.

Shape and Texture What makes graphene so strong? The property of shape is part of graphene's super strength. If you look at a very magnified view of graphene, you see it looks like chicken wire. Each of these shapes works together to form a strong network. The **texture** of a sheet of graphene is smooth.

Someone can see through chicken wire by looking through the holes.

Graphene looks a lot like chicken wire.

texture

Texture describes the surface of any area made up of matter.

Mass and Volume Graphene does not have much **mass** . A layer large enough to cover a football field would have a mass of one gram, about the same mass as a teaspoon of sugar.

Do you remember how to measure **volume** ? What would be the volume of enough graphene to cover a football field? A football field is 110 m (360 ft) long and 49 m (160 ft) wide. The height of graphene is less than a nanometer. So if you multiplied length × width × height, you'd find a very small volume of graphene.

MASS AND VOLUME OF GRAPHENE

How much graphene does it take to cover a football field?

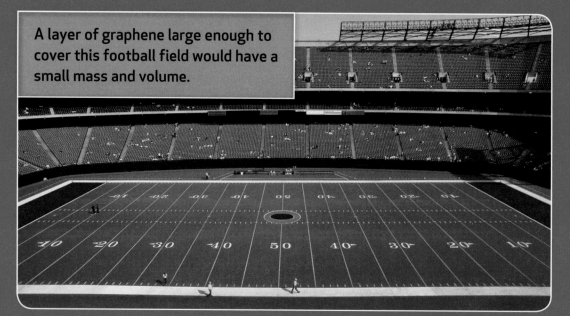

A layer of graphene large enough to cover this football field would have a small mass and volume.

mass

Mass is the amount of matter in an object.

volume

Volume is the amount of space matter takes up.

Graphene can store energy. It allows energy to flow very quickly, too. That means that graphene can be part of new technology. Scientists are excited about the many things they think graphene can do. Scientists think that graphene could help them make electronic paper like computer screens. The screens would be so thin, you could roll them up and carry them. Graphene might help scientists create faster computers and better cell phones. Graphene has an amazing future!

Graphene can be used in solar power cells. Graphene is so thin, it lets light through.

CHAPTER 1 — SHARE AND COMPARE

Turn and Talk How can properties help us describe graphene? Form a complete answer to this question together with a partner.

Read Select two pages in this section. Practice reading the pages. Then read them aloud to a partner. Talk about why the pages are so interesting.

my SCIENCE notebook

Write Write a conclusion that tells the important ideas you learned about graphene. State what you think is the Big Idea of this section. Share what you wrote with a classmate. Did your classmate write about the properties of graphene and what makes it unique?

my SCIENCE notebook

Draw Suppose you magnified graphene even closer than the magnifications shown in this book. Draw what you think this close-up view would look like. Combine your drawings with those of your classmates to create a graphene gallery.

CHAPTER

2

WHAT ARE STATES OF MATTER?

If you look carefully at the photograph, you'll see something you see every day: water! Water is made of matter, just like everything else you can see around you. Liquid water flows in this creek. Solid water, or ice, covers some of the grasses near the stream. Each form of water is a state. What states of water do you see every day?

TECHTREK
myNGconnect.com

Fog gathers above this creek in California. The water in the creek is much warmer than the snow-covered banks.

38   39

After reading Chapter 2, you will be able to:

- Recognize that all objects are made of matter and that matter takes up space and has mass. **STATES OF MATTER**

- Describe, classify, and explain the properties of solids, liquids, and gases and recognize that water can exist in all three states. **STATES OF MATTER**

- Recognize that air is a substance that surrounds us, has mass, and takes up space.
  STATES OF MATTER

- Understand that the properties of an object are dependent on the conditions of the present surroundings in which the object exists, such as temperature.
  MATTER CHANGES STATE

- Recognize that heating and cooling (temperature change) may cause changes in the properties of materials such as water, including phase changes in the state of matter.
  MATTER CHANGES STATE

- Measure and compare the temperature of water when it exists as a solid to its temperature when it exists as a liquid. **MATTER CHANGES STATE**

- Science in a Snap! Understand that the properties of an object are dependent on the conditions of the present surroundings in which the object exists, such as temperature.
  MATTER CHANGES STATE

WHAT ARE STATES

If you look carefully at the photograph, you'll see something you see every day: water! Water is made of matter, just like everything else you can see around you. Liquid water flows in this creek. Solid water, or ice, covers some of the grasses near the stream. Each form of water is a state. What states of water do you see every day?

OF MATTER?

TECHTREK
myNGconnect.com

Student eEdition

Vocabulary Games

Digital Library

Enrichment Activities

Fog gathers above this creek in California. The water in the creek is much warmer than the snow-covered banks.

SCIENCE VOCABULARY

states of matter
(STĀTS UV MA-ter)

States of matter are the forms in which a material can exist. (p. 42)

Solid ice is one state of matter of water.

solid (SO-lid)

A **solid** is matter that keeps its own shape. (p. 43)

This ice cube is a solid.

liquid (LI-kwid)

A **liquid** is matter that takes the shape of its container. (p. 44)

Liquid water takes the shape of the glass and the vase.

my Science Vocabulary

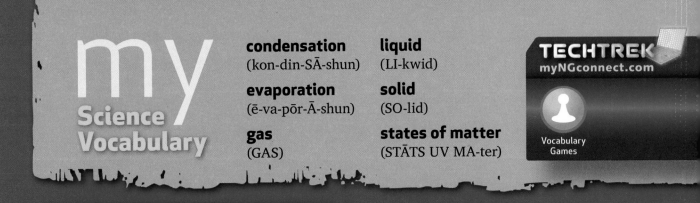

condensation
(kon-din-SĀ-shun)

evaporation
(ē-va-pōr-Ā-shun)

gas
(GAS)

liquid
(LI-kwid)

solid
(SO-lid)

states of matter
(STĀTS UV MA-ter)

TECHTREK
myNGconnect.com

Vocabulary
Games

gas (GAS)

A **gas** is matter that spreads to fill a space. (p. 45)

The gas in these balloons makes them float in the air.

evaporation
(ē-va-pōr-Ā-shun)

Evaporation is the change from a liquid to a gas. (p. 48)

Evaporation causes the water in a puddle to disappear from the road.

condensation
(kon-din-SĀ-shun)

Condensation is the change from a gas to a liquid. (p. 50)

The condensation of water vapor in the air formed the dew on these petals.

States of Matter

Take a close look at the photo. What do you see? You see matter! Everything around you is made of matter. Matter has mass and takes up space. Each piece of matter takes up its own space. That's why only one fishing pole, for example, can be in the same place at the same time.

Matter can have different forms. Water, for example, can flow. When it's frozen, it's hard. There is even water in the air you cannot see. Water can be a solid, liquid, or a gas. Each form of water is called a state. **States of matter** are the forms in which a material can exist.

The water in this stream is always warm. People can fish there during the cold winter because the water does not freeze.

Solid The **solid** form of water is ice. What makes a solid a solid? A solid is matter that keeps its own shape. If you take ice out of an ice cube tray and put it in a glass, it still keeps its shape as long as it is cold.

If you move ice from an ice cube tray to a glass, its shape stays the same.

Just like solid water, these blocks are solid. The blocks keep their shapes whether they are alone or in a stack.

A bowling ball is a solid. It always has the same shape and volume.

Ice cube trays hold frozen water we can use in many ways.

Liquid You know what happens when you knock over a glass of water. It forms a puddle. The water in the puddle has a different shape than the water in the glass had. That is because water is a **liquid** . A liquid is matter that takes the shape of its container. That liquid may change shape, but the amount of that liquid is the same no matter what container the liquid is in.

TECHTREK
myNGconnect.com

Enrichment Activities

Much of Earth is covered with liquid water.

TECHTREK
myNGconnect.com

Digital Library

Does the amount of water look different? Believe it or not, each container has the same amount of water in it. The shape of the water changed to fit the container.

Gas If you have ever watched a pot of water heat up on the stove, you may have noticed bubbles in the water. Those bubbles show that a **gas** is forming. A gas is matter that spreads to fill a space. The gas form of water is water vapor. Like most gases, water vapor is invisible. Air is a gas, too. Even though you cannot see air, it is still matter. It has a small amount of mass, and it takes up space.

Helium is a gas that fills some balloons. Helium makes balloons float in the air.

Before You Move On

1. In what three states can you find matter?
2. What is the name of the solid form of water?
3. Apply Is honey a solid, liquid, or gas? How do you know?

Matter Changes State

Walking on water is impossible, right? If you have ever skated on ice, though, you have almost done just that! What state is the water in the photo of the ice skaters?

You can find water in all three states in your kitchen. You can even change it from one state to another. Changes of state can happen when the temperature changes.

Skaters enjoy a cool day on an outdoor ice rink in London, England.

Freezing and Melting Ice forms when water freezes. Freezing is the change from a liquid to a solid. Freezing happens at low temperatures. It happens when water cools to its freezing point, which is 0°C (32°F).

Melting is the opposite of freezing. Melting is the change from a solid to a liquid. When the temperature rises above 0°C, ice changes to water. Other substances melt, too. If you have ever heated butter in a pan on the stove, you've probably seen a solid melt.

When the temperature rises to 0°C and climbs higher, icicles drip, get smaller, and soon disappear.

Evaporation On a humid day, the air feels wet. You cannot see it, but there is water in the air. Water that is a gas is called water vapor. It gets in the air through evaporation. Evaporation is the process in which water changes from a liquid to a gas. Evaporation can take place at many temperatures. It takes heat energy to cause evaporation, though.

The liquid water in the puddle evaporated into the air. The gas form of air, called water vapor, is invisible.

Boiling Look at the pot of water boiling on the stove. What do you observe? You probably notice the bubbles that show that water is changing from a liquid to a gas. Water makes this change from a liquid to a gas at 100°C (212°F). That temperature is the boiling point of water. Other liquids have different boiling points.

TECHTREK
myNGconnect.com

What changes can you see in water as it begins to boil?

Digital Library

100°C (212°F) ⟶

Condensation See the white fluffy clouds in the photograph? How did they form? Clouds form because of water. Air contains invisible water in the form of a gas called water vapor. The air high in the sky is cool. As the air cools, tiny drops of water form. Condensation is the change from a gas to a liquid. When water condenses high in the air, it sticks to pieces of dust and floats in the air. When those pieces of dust covered with water come together, they make a cloud you can see.

Clouds form when the temperature of the air changes.

Condensation doesn't happen only high in the air. On the ground, dew can form on grass or flowers. Dew forms when water vapor in the air condenses. If you have a glass of cold lemonade on a warm day, you may notice drops of water on the glass. The drops of water are condensation. The warmer air causes the water vapor in the air to condense on the cold drinking glass.

Dew drops formed on these flower petals. They form when water vapor in the air cools.

Before You Move On

1. Name two actions that cause matter to change from liquid to gas.
2. What changes of state can happen when temperatures get colder?
3. **Apply** After a rainstorm, the sidewalk dries out. What happens? Where does the water go?

Measure Temperature

You already know that water changes state when the temperature changes. But how can you measure the temperature? You can use a special tool, a thermometer, to measure the temperature.

Look at the thermometer floating in the pool. The thermometer measures the temperature of liquid water. You could read the thermometer to find out if the water is warm enough for swimming.

A thermometer measures the temperature of this bubbly liquid cooking on the stove.

This floating thermometer measures the temperature of the liquid water in the pool.

You can measure the temperature of solids, too. Meat is a solid. Cooks use thermometers to help them decide when meat is finished cooking. When the meat reaches the correct temperature, the cook knows that it is safe for people to eat.

Science in a Snap! Find the Temperature

Look at the oil, corn syrup, and water. Describe them. How are they the same? How are they different?

What is the temperature of each liquid?

Put a thermometer in each cup.

Before You Move On

1. Describe how you would measure the temperature of the soil outside your school.
2. Compare the temperature of water in its liquid state with the temperature of water in its solid state.
3. **Infer** An ice cube falls on the ground outside. A thermometer shows that the outside air temperature is below 0°C. What will happen to the ice cube?

FOG FENCES
IN CENTRAL AMERICA

In some places in Guatemala, it is hard to get fresh water. It does not rain very often. There are few wells. It is hard to build pipes to bring in water from cities that are far away.

The fog fence collects water that helps this girl and her family drink, cook, bathe, and grow food.

The people who live in Guatemala still need water. They need it to drink and cook. They need it to feed their animals and grow plants for food. One way to get water is to build a fog fence. A fog fence uses condensation to collect water.

Fog is a cloud of water droplets floating in the air near the ground. A fog fence is a large plastic net. When wind blows fog through the net, the net gets wet. The water rolls or trickles down into pipes. A net that is 8 meters (26.2 feet) high and 50 meters (164 feet) long collects about 200 liters (52.8 gallons) each day. That amount would fill about four bathtubs.

Fog forms when water in the air condenses to form a low cloud.

Drops of water from fog collect on the plastic strands of the fog nets.

Water from the fog fence collects in a large tank.

Matter exists in three states: solid, liquid, and gas. Matter changes state because of changes in temperature. Freezing and condensation happen when water is cooled. Melting, boiling, and evaporation happen when water is heated.

Big Idea Matter exists in three states. Changes in state are caused by changes in temperature.

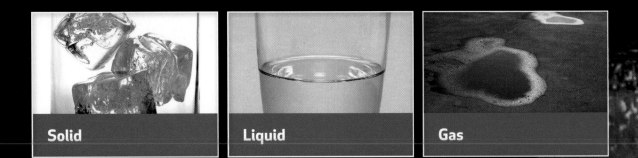

Solid | Liquid | Gas

Vocabulary Review

Match the following terms with the correct definition.

A. condensation

B. evaporation

C. gas

D. states of matter

E. liquid

F. solid

1. Matter that keeps its own shape

2. Matter that spreads to fill a space

3. Matter that takes the shape of its container

4. The change from a gas to a liquid

5. The change from a liquid to a gas

6. The forms in which a material can exist

Big Idea Review

1. **Define** Give definitions for these terms: solid, liquid, gas. Tell what water is like in each of these states.

2. **Explain** What happens to the shape of a liquid when it is poured into a container with a different shape?

3. **Describe** How is evaporation part of boiling?

4. **Cause and Effect** What would happen to a jar of water vapor if you put it in the refrigerator?

5. **Evaluate** A friend says to you, "Matter is very useful in its liquid and solid states, but not in its gas state." How would you respond?

6. **Infer** What kinds of places can use fog fences?

Write About Matter

Describe What is happening in this photograph? What states of matter are there? What change is taking place? How is temperature important?

CHAPTER 2 PHYSICAL SCIENCE EXPERT: ICE SCULPTOR

What Does an Ice Sculptor Do?

Some artists make sculptures out of stone. Others use ice! Steve Brice wins contests for carving beautiful shapes out of solid ice. He works in Chena Hot Springs in the northern state of Alaska. The temperature is below freezing much of the year there. So, his sculptures last for months outside without melting.

Steve Brice

Colorful lights brighten up the sculptures in Steve Brice's gallery.

Steve Brice invented this special tool
for carving details in ice. He uses it
to put writing on his sculptures.

Ice is hard like stone and wood. An ice sculptor uses many of
the same tools to carve ice that other sculptors use to shape
other solids. Brice uses chain saws, sanders, knives, and drills
to shape the ice.

As long as it stays cold, the finished sculptures keep their
shape. Every year, the sculptures melt in the spring. Brice
makes new ones when it freezes again in winter.

Steve Brice uses a special
tool to smooth the face of
one of his ice sculptures.

BECOME AN EXPERT

Sweden: Ice Hotel Construction

Buildings have to keep their shapes. They cannot bend or flow. This is why they have to be made out of **solid** things. Did you know that a building can be made out of water? A hotel in the far northern part of the country of Sweden is made out of water in its solid state—ice!

The temperature has to be well below 0°C (32°F) to keep the ice—the hotel—solidly frozen.

solid

A **solid** is matter that keeps its own shape.

TECHTREK
myNGconnect.com

Student
eEdition

Digital
Library

Freezing Ice Blocks

In some places, it gets very cold in the winter. When it gets below 0°C (32°F), water freezes. In cold places such as northern Sweden, the **liquid** water in lakes and rivers can freeze. When a river freezes, the water at the top turns solid. The people who build the ice hotel use large blocks of this frozen river water.

Workers haul away large blocks of ice. They move ice from the river to the place where they will build the hotel.

liquid

A **liquid** is matter that takes the shape of its container.

61

Building with Ice and Snow

Some parts of the ice hotel are made
with packed snow. Snow is a powdery
solid. When it is packed down, it makes
a hard shape. Builders of the ice hotel
use packed snow to make the walls and
ceilings of the ice hotel.

1. Forms are set into place.
These forms give the snow
the right shape.

2. A snow cannon shoots snow
onto the forms. It is packed
into a solid shape on the form.

3. When the packed snow has the right shape, the form is taken out. It can be used again to build another part of the hotel.

HOW DOES BUILDING WITH ICE WORK?

Winter temperatures drop, and the river freezes.

Blocks of ice are cut from the frozen river.

The blocks are shaped and stacked to make pillars, beds, and so on.

Living in Ice

To keep the ice frozen, it must stay cold inside the hotel. The rooms are usually five degrees colder than 0°C. Outside, it can be 20 or 30 degrees colder than that temperature.

Even the ceiling light fixture is made of ice.

The columns and the furniture are carved from frozen blocks of ice taken from the nearby river.

Seeing Your Breath A guest at the ice hotel might be able to see her breath. When it is cold, sometimes it looks like clouds are coming out of your mouth when you breathe. The puffs are not a **gas** . They are tiny drops of water. They formed when warm air that you breathed out cooled. In other words, **condensation** occurred.

TECHTREK
myNGconnect.com

Digital Library

Guests sleep on beds made of ice and covered with furs.

It is cold enough inside that a guest must wear her jacket.

gas

A **gas** is matter that spreads to fill a space.

condensation

Condensation is the change from a gas to a liquid.

65

The Hotel Melts

When spring comes, it gets warm again. Soon, the temperature gets above 0°C. The ice of the hotel cannot stay solid. It begins to melt. As the blocks melt, liquid water forms. It drips from the ceiling. It collects in puddles. The hotel is completely gone!

WHERE DID THE ICE HOTEL GO?

Water in nature can change between all three states of matter.

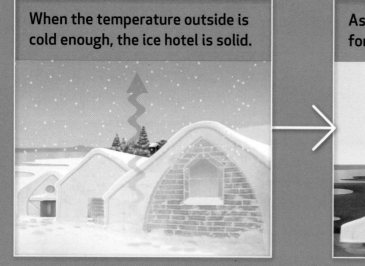

When the temperature outside is cold enough, the ice hotel is solid.

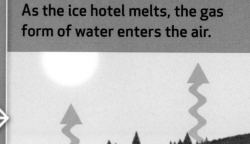

As the ice hotel melts, the gas form of water enters the air.

Gas Again   In just two months, all of the ice that made up the hotel melts. The water from the melting ice runs into the river. **Evaporation** occurs, and some water becomes water vapor. The ice hotel's water has changed into all three **states of matter** . Next winter, the water in the river will freeze again. It can be used to build a new ice hotel.

When the ice melts, the hotel slowly disappears.

evaporation
Evaporation is the change from a liquid to gas.

states of matter
States of matter are the forms in which a material can exist.

CHAPTER 2
SHARE AND COMPARE

Turn and Talk How do states of matter affect the Ice Hotel? Form a complete answer to this question together with a partner.

Read Select two pages in this section. Practice reading the pages. Then read them aloud to a partner. Talk about why the pages are interesting.

my **SCIENCE** notebook

Write Write a conclusion that tells the important ideas you learned about constructing an ice hotel. State what you think is the Big Idea of this section. Share what you wrote with a classmate. Did your classmate make the connection between the states of water and how they are important to constructing the hotel?

my **SCIENCE** notebook

Draw Draw a picture of a stage in the making or the melting of the ice hotel. Label the states of matter and the changes of state that are happening. Combine your drawings with those of your classmates to create a guidebook for the Ice Hotel.

CHAPTER
3

HOW DOES FORCE CHANGE MOTION?

A wagon moves when you pull on it. A shopping cart goes forward when you push on it. A ball falls to the ground when you drop it. Why does it fall? It is being pulled to Earth. Pushes and pulls make objects move. They also change the way objects move.

This scarab beetle pushes an object and makes it move.

70

71

After reading Chapter 3, you will be able to:

- Describe a force as a push or a pull, and recognize that some forces will only make objects move if they are touched. **MOTION, GRAVITY, MAGNETISM**

- Describe the position, direction, and motion of objects, and how the motion of objects changes by speeding up or slowing down. **POSITION, MOTION**

- Recognize that an object's speed depends on the time it takes to go a certain distance. **MOTION**

- Explain how the mass of objects and the strength of forces affect motion. **MOTION, GRAVITY, MAGNETISM**

- Explain that friction slows the motion of objects. **MOTION**

- Explain that gravity is a force that pulls objects toward Earth. **GRAVITY**

- Explain that magnetism is a force that pulls on some objects without touching them. **MAGNETISM**

- ⬤ Science in a Snap! Recognize that an object's speed depends on the time it takes to go a certain distance. **MOTION**

HOW DOES

FORCE

A wagon moves when you pull on it. A shopping cart goes forward when you push on it. A ball falls to the ground when you drop it. Why does it fall? It is being pulled to Earth. Pushes and pulls make objects move. They also change the way objects move.

TECHTREK
myNGconnect.com

Student
eEdition

Vocabulary
Games

Digital
Library

Enrichment
Activities

CHANGE MOTION?

This scarab beetle pushes an object and makes it move.

SCIENCE VOCABULARY

motion (MŌ-shun)

When an object is moving, it is in **motion** . (p. 76)

The skier is moving from one place to another place.

speed (SPĒD)

Speed is the distance an object moves in a period of time. (p. 78)

Speed depends on how far an object goes. It also depends on how long it takes to move that far.

force (FORS)

A **force** is a push or a pull. (p. 80)

The weightlifter moves the barbell by pulling on it.

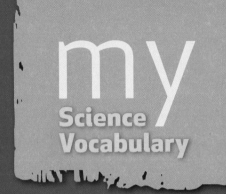

my Science Vocabulary

force (FORS)

friction (FRIK-shun)

gravity (GRA-vi-tē)

magnetism (MAG-na-ti-sum)

motion (MŌ-shun)

speed (SPĒD)

TECHTREK
myNGconnect.com

Vocabulary Games

friction (FRIK-shun)

Friction is a force that acts when two surfaces rub together. (p. 84)

The penguins can slide on the ice because there is not much friction.

gravity (GRA-vi-tē)

Earth's **gravity** is a force that pulls things to the center of Earth. (p. 86)

Gravity will pull the ball to the ground.

magnetism (MAG-na-ti-sum)

Magnetism is a force between magnets and the objects magnets attract. (p. 90)

Magnets do not pull on plastic objects.

73

Position

Look at the photo. Where are the acrobats? You can use the balls to describe the positions of the acrobats. The position of the acrobats is above the balls. Position describes where an object is. You always compare one object to another object when you describe position.

The position of the balls is under the acrobats.

There are many ways you can describe position. Look at the photo of the three friends. The boy is in the middle. He is holding a giant pumpkin in front of his body. One girl is wearing a hat on top of her head. She is holding a pumpkin beside her head. The other girl is holding a pumpkin over her head. You compare the location of objects using phrases that describe position.

The position of each pumpkin is different.

Before You Move On

1. What is position?
2. What are some words you can use to describe the position of objects?
3. **Apply** A bird is flying over a cat. Describe the position of the cat by comparing its position with that of the bird.

Motion

The skier in the picture is in motion . Things that are in motion change position. The skier starts at the top of the hill. He moves farther down the hill. His position changes as he moves down the hill.

The skier changes position quickly as he goes down the mountain.

One way you describe motion is by describing direction. Direction is the path an object takes. Look at the picture. Some of these cars are going straight. Others might turn left or right. *Left*, *right*, and *straight* are words that describe direction. *North*, *south*, *east*, *west*, *up*, and *down* are also direction words.

The cab drivers and the people walking on the street are moving in different directions.

Speed Motion can be fast or slow. **Speed** is the rate at which an object changes position. It tells you how fast an object is moving. A race car moving at a high speed changes position faster than a race car moving at a slower speed. The car that completes the race in the least amount of time has the greatest speed. The car that completes the race in the greatest amount of time has the slowest speed.

TECHTREK
myNGconnect.com

Digital Library

This race car's speed is over 280 km (175 miles) per hour!

Put a piece of tape at your starting position. Starting at the piece of tape, measure ten meters with a partner. Put another piece of tape at the end of ten meters.

Walk ten meters. Time your partner and have your partner time you. Observe your speed and your partner's speed closely.

How were your speeds different? Who had the greater speed?

Before You Move On

1. What is motion?
2. How are speed and motion related?
3. **Apply** You are watching two people running in a race. How do you know which person is moving at a greater speed?

Force

Look at the photos. What do you observe about the woman and the barbell? How does she get the barbell to move?

The woman is pulling on the barbell. She must use **force** to move it. A force is a push or a pull. Forces can change the position of objects. You pull objects toward you. The weightlifter will move the barbell as she pulls it.

The weightlifter pulls on the barbell. The barbell moves because of the pull.

Forces cause motion. Objects stay in one place until a force makes them move. More force is needed to move objects that have more mass. The weightlifter uses a large amount of force to push the barbell over her head. You use a small force to push a pillow above your head.

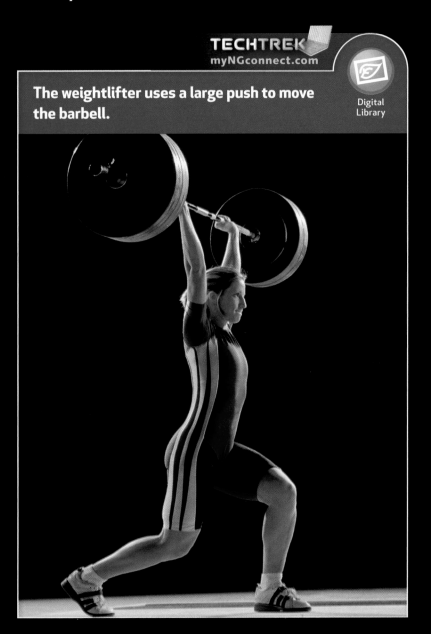

The weightlifter uses a large push to move the barbell.

Friction Suppose you are riding your bike. You stop pedaling. Your bike will start to slow down. Eventually it will stop. It stops because of the **friction** between the tires and ground. Friction is a force between two objects that rub together. It acts against motion. It slows moving objects. Your bike tires and the ground rub together. This helps the bike stop.

TECHTREK
myNGconnect.com

Digital Library

There is friction between the bike wheels and the ground.

Look at the picture of the penguins. What do you observe about their motion? What does this tell you about the force between the penguins and the ice?

The penguins are sliding on the ice. There is only a little friction between their bellies and the ice. They would not be able to slide on a road. The road has a rougher surface, so there is more friction.

Penguins can slide on the smooth ice. There is not much friction.

Before You Move On

1. How is kicking a ball an example of a force?
2. Does it take more force to move an apple or a watermelon? Explain how you know.
3. **Infer** How does the friction between car tires and pavement compare to the friction between car tires and ice?

Gravity

What is this football player doing? He is kicking the ball. The ball goes high in the air. But then what happens? The ball falls back to the ground. Why? Gravity pulls it down. Earth's gravity is a force that pulls things to the center of Earth.

Gravity is one of many different forces that does not have to touch objects to have an effect on their motion. Gravity pulls on objects without touching them.

Gravity will pull this ball back to the ground after the player kicks it into the air.

Gravity pulls on everything on Earth. Try this simple test. Hold your hands over your head. What happens after a few minutes? Your arms start to feel tired. Why did your arms get tired? Gravity was pulling down on them. Gravity made your arms feel tired and heavy. It feels good to let your arms down and relax.

Gravity pulls everything toward the center of Earth, even you!

Has a doctor ever measured your weight? Weight depends on gravity. It also depends on the mass of objects. The pull of gravity is stronger on objects with more mass. This means objects with more mass weigh more. A stack of books feels heavier than one book. The pull of gravity is greater on the stack of books. The weight of the books is the measure of the pull of Earth's gravity on the books.

Weight measures the pull of gravity.

Before You Move On

1. What is gravity?
2. What does weight measure?
3. **Apply** One grocery bag is filled with jars of peanut butter and jelly. Another grocery bag is filled with bread. Why does the second grocery bag weigh less than the first one?

Magnetism

Magnetism is another type of force. It also pulls on objects without touching them. Magnetism is a force between magnets and some metal objects. Magnets pull on some metals, such as iron. The force of magnetism does not pull everything. It does not pull on plastic or wood.

The big magnet can pick up metal garbage without touching it.

TECHTREK
myNGconnect.com

Enrichment Activities

Magnets pull on metal objects.
They do not pull on plastic objects.

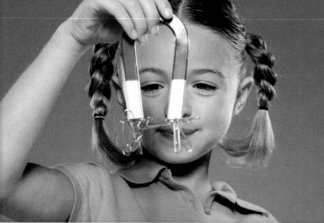

Magnets can pull on objects through other materials. Small magnets can pull through paper. That is how they can hold a piece of paper on the refrigerator. The magnets are not attracted to the paper. They pull on the door through the paper. The force of magnetism also depends on how far the magnet is from the object it pulls on. The force gets weaker as the magnet gets farther away.

Magnets pull on the refrigerator door, not the paper.

Look at the picture of the magnet. Why are the iron filings moving toward certain places on the magnet more than others? The iron filings are attracted to the poles of the magnet. All magnets have two poles. One end of the magnet is the north pole. The other end is the south pole. The force of the magnet is strongest at the poles. It gets weaker farther away from the poles.

The pull of the magnet is strongest at its poles.

pole

pole

Poles that are the same repel, or push away from each other. A north pole of one magnet pushes away from the north pole of another magnet. Two south poles also push away from each other. Opposite poles attract, or pull toward each other. A north pole and a south pole pull together.

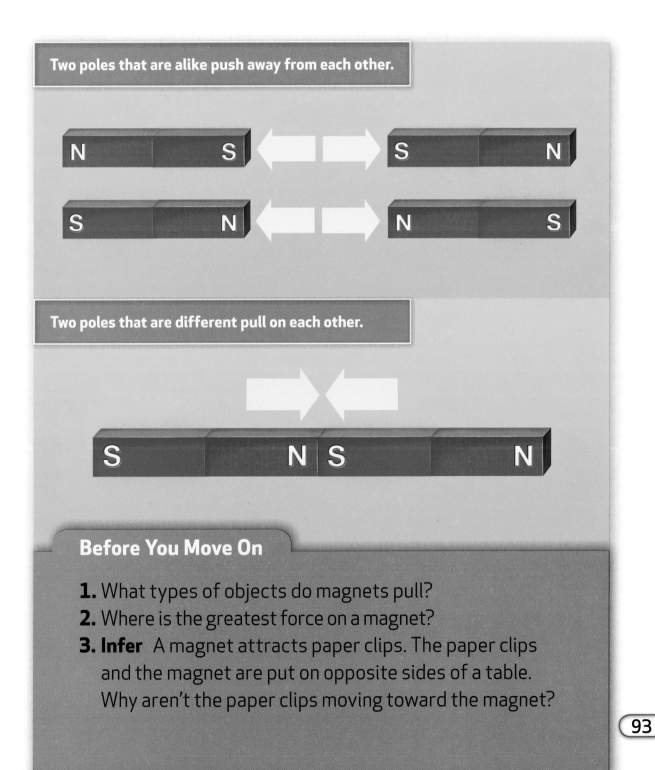

Two poles that are alike push away from each other.

Two poles that are different pull on each other.

Before You Move On

1. What types of objects do magnets pull?
2. Where is the greatest force on a magnet?
3. **Infer** A magnet attracts paper clips. The paper clips and the magnet are put on opposite sides of a table. Why aren't the paper clips moving toward the magnet?

PARACHUTES
FLOATING THROUGH THE AIR

Most people cannot imagine jumping out of an airplane. Gravity pulls you to Earth. You go faster and faster. The ground gets closer and closer. Luckily, skydivers have parachutes. Parachutes slow them down before they hit the ground.

Parachutes slow a skydiver's fall.

Modern parachutes are made of nylon. Nylon is a strong material. It is also thin and light. Parachutes have a large surface. Air pushes on this surface when they open. This pushing is a kind of friction. It resists the force of gravity. The parachute slows the motion toward the ground.

Parachutes have many uses other than skydiving. Some planes use them to slow down when they land. Parachutes are also used to drop supplies from planes.

This is an early parachute design.

Fast planes use parachutes to help slow down after landing.

Objects in motion change their position. Speed and direction are ways to describe motion. A force is a push or a pull. Forces change the motion of objects. Friction is a force that works against the motion of objects. It affects objects that rub together. Earth's gravity is force that pulls objects to the center of Earth. Magnets can pull some metals without touching them.

Big Idea Forces are pushes and pulls that can change the motion of objects.

Pull

OR

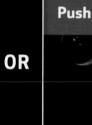
Push

━━━ **FORCE**

Vocabulary Review

Match the following terms with the correct definition.

A. force

B. friction

C. gravity

D. magnetism

E. motion

F. speed

1. The distance an object moves in a period of time

2. A force that pulls things to the center of Earth

3. A force that acts when two surfaces rub together

4. A force between magnets and the objects magnets attract

5. A push or a pull

6. Describes an object that is moving

Big Idea Review

1. **Recall** What is motion?

2. **Describe** How could you describe your position in the classroom?

3. **Explain** What causes the motion of objects to change?

4. **Compare and Contrast** How are gravity and magnetism alike and different?

5. **Infer** Soccer players wear special shoes. The bottoms of the shoes have large bumps that stick into the ground. Why are the shoes made like this?

6. **Analyze** Bob and Kate are riding skateboards. Bob rides 10 meters in 10 seconds. Kate rides 20 meters in 10 seconds. Who has the greater speed?

Write About Forces and Motion

Explain What is happening to the magnets in the diagram? Why is this happening?

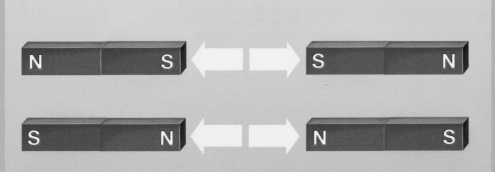

CHAPTER 3

PHYSICAL SCIENCE EXPERT: MATERIALS SCIENTIST

A inissa Ramirez is an Associate Professor of Mechanical Engineering at Yale University. She is a scientist that works with other scientists to study the properties of different materials and the ways those materials respond to different forces.

TECHTREK
myNGconnect.com

Digital Library

Ainissa Ramirez

Q: What do you study?

My research group and I are looking at how materials behave when they are really really small—less then 1/100,000 the thickness of your hair. Since these objects are too small to "see," we use a machine, called an Atomic Force Microscope, which "feels" atoms with a very sharp tip. When the tip runs over the surface, forces between the tip and the atoms are fed into a computer and become images of atoms that we can examine.

Q: Why is what you study important?

These studies are important because properties of materials change when you change the size of the material. And, in order to use these materials in future appliances, like cell phones and video games, we must map how they behave.

We explore all the properties of materials and map them out for future scientists. That is what science is about—scientists building off of each other's work because there is a lot to learn.

Q: What is your favorite thing about being a materials scientist?

I am still amazed at how smart atoms are. The study of atoms and how they interact is fascinating and there are plenty of interesting things about them that we still do not know. Being a scientist is a very rewarding career. If you are interested in materials or atoms or forces—Jump in!

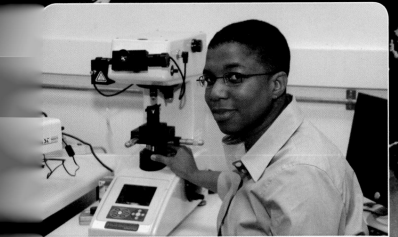

Ainissa uses a hardness tester to press on materials with a force to see how strong the materials are.

This image shows a close-up of the shape of the surface of a piece of polished aluminum under a microscope.

BECOME AN EXPERT

High-Speed Trains

Trains were once the fastest way to get around. In 1869, a train took six whole days to cross the country. The **speed** that train could move would seem slow now. Today, it only takes six hours to cross the country in a plane!

Steam engines were once used to make trains move.

speed
Speed is the distance an object moves in a period of time.

TECHTREK
myNGconnect.com

Student
eEdition

Digital
Library

Train engines are in special cars. These cars are at the front or back of the train. The engine provides the **force** that moves the train. A large force is needed to move a train. Early trains had steam engines. In the 1930s, new train engines that used diesel fuel were used. Many trains still use these types of engines today. Other trains use electricity.

People have ridden trains to get where they need to go for many years.

force
A **force** is a push or a pull.

Bullet Trains Some trains move very fast. They are called bullet trains. Some of these trains have a top speed that is over 320 kilometers (200 miles) per hour! Most of these trains have electric engines. How does a train get electricity while it is moving? There are wires above the train tracks. The trains have special parts on top. These parts allow the train to get electricity from the wires. In this way, the train can get electricity while it is in **motion**.

TECHTREK
myNGconnect.com

Digital Library

This train gets electricity from the wires above the tracks.

motion

When an object is moving, it is in **motion.**

The Acela train is the fastest train in the United States. It takes people between Boston and New York City. It also goes between Washington, D.C., and New York City. Its top speed is about 240 kilometers (150 miles) per hour. It only goes this fast where the train track is straight. It slows down when the track changes direction.

Bullet trains help people get from place to place faster.

Maglev Trains Some of the fastest trains in the world are in Germany. One of the trains, the Transrapid, can go up to 434 kilometers (271 miles) per hour. This train is different from electric trains. It doesn't have an engine car. It doesn't even have wheels! It glides on a thin cushion of air above the tracks. The train is a maglev train. The word "maglev" is short for "magnetic levitation." These trains are called maglev trains because the force of **magnetism** lifts them above the tracks and makes it seem like they are "floating."

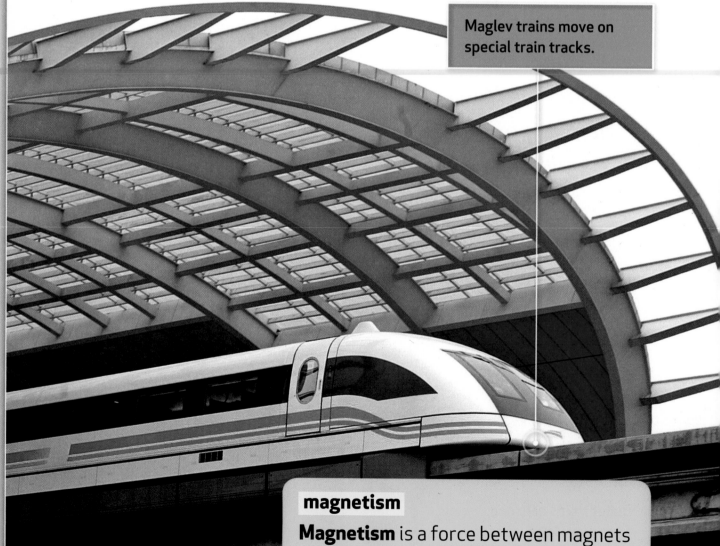

Maglev trains move on special train tracks.

magnetism
Magnetism is a force between magnets and the objects that magnets attract.

A train floating in the air might seem like magic. After all, trains are big. They are very heavy. The pull of **gravity** on the train is strong. Of course, it is not magic. The magnets keep the trains up off the track. The magnets' pull is greater than the pull of gravity. These magnets keep the maglev train riding safely on the thin cushion of air above the tracks.

HOW A MAGLEV TRAIN WORKS

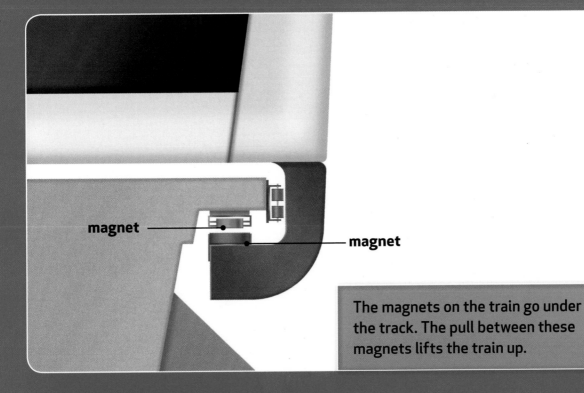

magnet

magnet

The magnets on the train go under the track. The pull between these magnets lifts the train up.

gravity
Earth's **gravity** is a force that pulls things to the center of Earth.

Maglev trains use special tracks that have coils. The coils are turned into special magnets that use electricity. The magnets are only turned on when the train is nearby. The magnets in the track push and pull on the magnets in the train. These pushes and pulls move the train forward and stop the train as well. **Friction** is not a problem for maglev trains. Why? The trains do not rub against the tracks. They float above the tracks. Without much friction, the trains can go very fast.

There is almost no friction between this maglev train and the track.

friction

Friction is a force that acts when two surfaces rub together.

The Future of Maglev Trains Maglev trains have a lot of good properties. They do not make noise. The ride from one place to another is very smooth. They are fast. But there's one problem—the cost of building tracks. Maglev train tracks are very expensive to build. People may want to spend the money to make cheaper methods of travel better. However, maglev trains will probably have a place in our high-speed future!

Maglev trains are quiet, fast, and ride smoothly.

CHAPTER 3

SHARE AND COMPARE

Turn and Talk What are the advantages and disadvantages of high-speed trains? Form a complete answer to this question together with a partner.

Read Select two pages in this section. Practice reading the pages. Then read them aloud to a partner. Talk about why the pages are interesting.

my SCIENCE notebook **Write** Write a conclusion that summarizes what you have learned about high-speed trains. In your conclusion, restate what you think is the Big Idea of this section. Share what you wrote with a classmate. Compare what each of you wrote. Did you recall how high-speed trains use different types of force?

my SCIENCE notebook **Draw** Draw a picture of a high-speed train. Label the parts of the train. Share your drawing with a classmate. Combine your drawing with those of your classmates to show the different ways that forces make high-speed trains speed up or slow down.

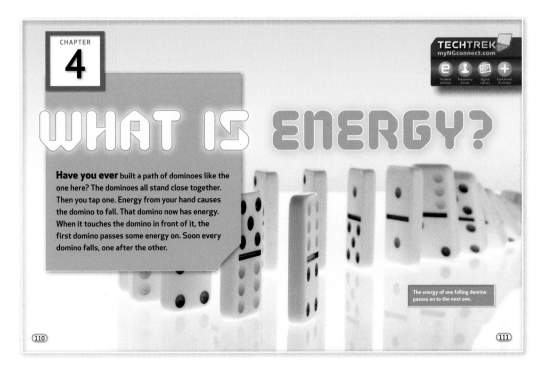

CHAPTER 4

WHAT IS ENERGY?

Have you ever built a path of dominoes like the one here? The dominoes all stand close together. Then you tap one. Energy from your hand causes the domino to fall. That domino now has energy. When it touches the domino in front of it, the first domino passes some energy on. Soon every domino falls, one after the other.

TECHTREK
myNGconnect.com

The energy of one falling domino passes on to the next one.

110   111

After reading Chapter 4, you will be able to:

- Identify and describe different forms of energy, including mechanical, sound, electrical, and heat. **MECHANICAL ENERGY, SOUND, ELECTRICAL ENERGY, HEAT**

- Recognize that energy has the ability to cause motion or create change. **ENERGY AND WORK, MECHANICAL ENERGY, ELECTRICAL ENERGY, HEAT**

- Identify and describe that vibrations are the source of sound energy. **SOUND**

- Distinguish materials that heat flows easily through and those it does not. **HEAT**

- Demonstrate that rubbing objects together produces heat. **HEAT**

- **Science in a Snap!** Demonstrate that rubbing objects together produces heat. **HEAT**

WHAT IS

Have you ever built a path of dominoes like the one here? The dominoes all stand close together. Then you tap one. Energy from your hand causes the domino to fall. That domino now has energy. When it touches the domino in front of it, the first domino passes some energy on. Soon every domino falls, one after the other.

ENERGY?

The energy of one falling domino passes on to the next one.

SCIENCE VOCABULARY

energy (EN-ur-jē)

Energy is the ability to do work or cause a change. (p. 114)

> These children use energy to play on a playground.

mechanical energy (mi-CAN-i-kul E-nur-jē)

The **mechanical energy** of an object is its stored energy plus its energy of motion. (p. 117)

> This skier has mechanical energy.

sound (SOWND)

Sound is energy that can be heard. (p. 120)

> This musician creates sound as he beats a drum.

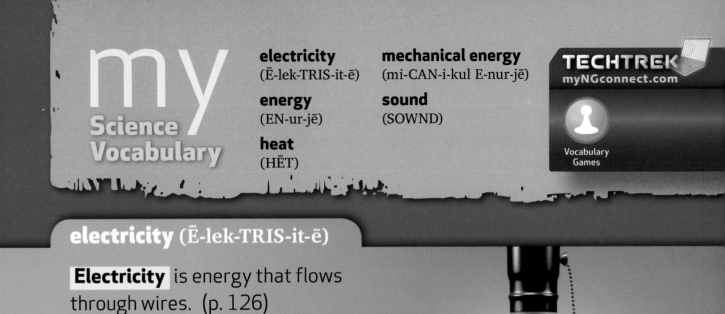

my Science Vocabulary

electricity
(Ē-lek-TRIS-it-ē)

energy
(EN-ur-jē)

heat
(HĒT)

mechanical energy
(mi-CAN-i-kul E-nur-jē)

sound
(SOWND)

TECHTREK
myNGconnect.com

Vocabulary
Games

electricity (Ē-lek-TRIS-it-ē)

Electricity is energy that flows through wires. (p. 126)

This light bulb uses electricity to light up a room.

heat (HĒT)

Heat is the flow of energy from a warmer object to a cooler object. (p. 130)

Heat flows from the warm soup to the cool spoon.

Mechanical Energy

Look at the picture below. The children ready to sled down the hill have a lot of stored energy. Stored energy is energy that is ready to be used.

These children have a lot of stored energy.

Look at the children sledding down the hill. Their stored energy has turned into a new type of energy, the energy of motion. All moving things have the energy of motion. The mechanical energy of each child is their stored energy plus their energy of motion.

The children have the energy of motion as they sled down the hill.

Let's look at another example of mechanical energy. A skier starts at the top of a mountain. She has stored energy. She is not moving. She does not have any energy of motion.

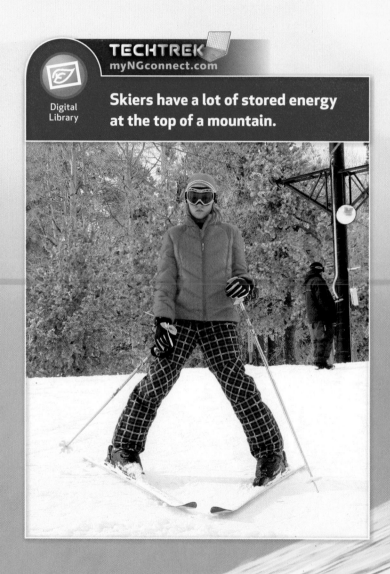

TECHTREK
myNGconnect.com

Digital
Library

Skiers have a lot of stored energy at the top of a mountain.

Look at this skier. He is in motion. His stored energy changes to energy of motion as he moves down the mountain. The skier's stored energy plus his energy of motion is his mechanical energy.

As the skier moves down the hill, his stored energy changes to energy of motion.

Before You Move On

1. What is mechanical energy?
2. How is stored energy different than energy of motion?
3. **Apply** You are running down a hill. How does your energy change?

Sound

Have you heard someone play a drum? When the stick hits the drum, the drum vibrates, or moves quickly back and forth. When the top of the drum vibrates, the air near the drum begins to vibrate too. These vibrations make sound. **Sound** is energy that can be heard!

This boy creates sound by hitting the skin of a drum.

You hear sound when sound travels to your ear. Look at your ear in a mirror. It is made to collect sound vibrations. The vibrations move to your ear, and you hear the sound.

Volume A jet engine is very loud when you are near it. It has a loud volume. Volume is how loud or soft a sound is. Strong vibrations make loud sounds. People working on the ground at an airport wear special headphones to protect their ears.

Without the headphones, the sound could damage this airport worker's hearing.

When you speak, the vocal cords inside your throat vibrate. When you shout, your vocal cords vibrate more. This makes your voice sound louder.

When you speak quietly, your vocal cords vibrate less. This makes your voice sound softer.

Pitch   Look at the animals on this page. What kinds of sounds do you think they make?

The kitten has thin, short vocal cords that vibrate quickly. It makes a high sound. The tiger has thick, long vocal cords that vibrate more slowly. It makes a low sound. Pitch is how high or low a sound is.

The kitten has a high-pitched meow.

The tiger has a low-pitched roar.

Look at the guitar. When the strings move, they vibrate. They make sound. A thick string vibrates slowly. The thick strings make low, deep sounds. This is a low pitch. A thin string can vibrate much faster. The thin strings make high sounds. This is a high pitch.

TECHTREK
myNGconnect.com

Thick strings make low-pitched sounds.
Thin strings make high-pitched sounds.

Enrichment
Activities

Before You Move On

1. What is sound?
2. Compare volume and pitch.
3. **Infer** Give an example of a sound with a loud volume and low pitch.

Electrical Energy

A DVD player needs energy to play movies. It uses **electricity**. Electricity is energy that flows through wires. You can watch the movies and hear the sound because of electricity.

The headphones change the electrical energy into sound you can hear.

Some DVD players plug into an outlet in the wall. You can also find DVD players that run on batteries. Batteries have stored energy. The DVD player changes the stored energy to electricity.

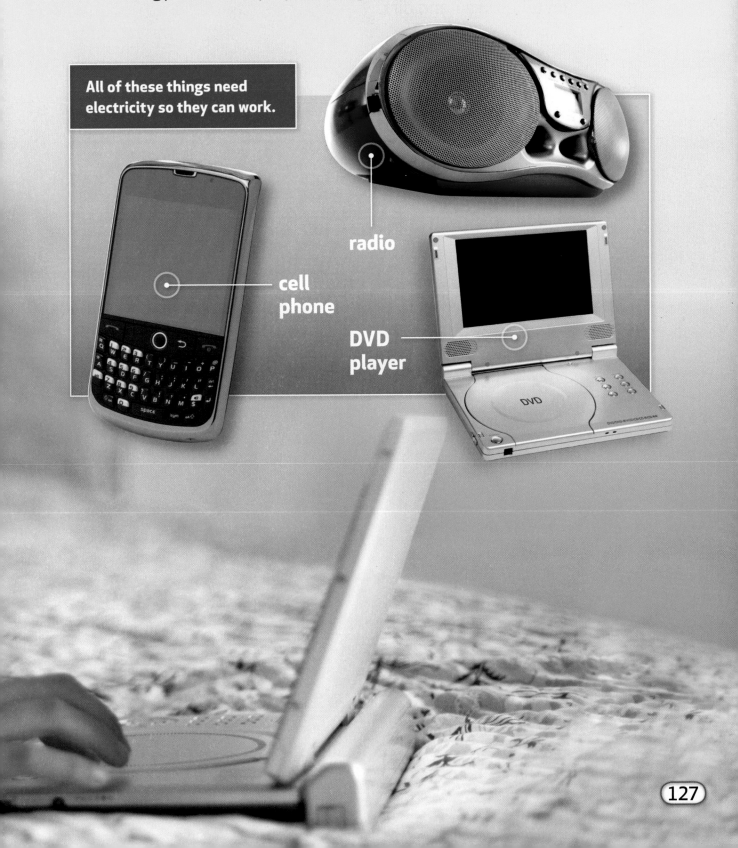

All of these things need electricity so they can work.

radio

cell phone

DVD player

This vacuum gets its power from electricity. Electricity is flowing through the wires inside the vacuum. The electricity becomes energy of motion. The wheels on the vacuum turn. The vacuum takes in air and dirt.

This vacuum cleaner uses electricity.

Electricity also flows through wires to make light bulbs shine. The light bulb turns the electricity into light. The light allows you to see all the objects around you.

Energy-saving light bulbs use less electricity.

Before You Move On

1. What is electricity?
2. How do electronic devices like DVD players get electricity?
3. **Apply** Choose something electric from your home. How does it change electricity into another form of energy?

Heat

This soup was just cooked. It is very hot. The metal spoon was in a drawer. It is much cooler than the soup. When the spoon is used to stir the soup, it quickly becomes hot. **Heat** moves from the soup to the spoon. Heat is the flow of energy from a warmer object to a cooler object.

The cool spoon heats up when you put it in hot soup.

Heat moves easily though some materials such as metal. The cookies bake on the hot metal cookie sheet. The woman wears cloth oven mitts to take the cookie sheet out of the oven. Heat does not move easily through cloth or plastic.

Thick, cloth oven mitts protect this woman's hand from the heat.

Look at the photo on this page. What do you think is happening to the rocks?

Two rocks are being rubbed together. They get so hot that a spark forms. Heat is made whenever two things rub together.

When these two rocks hit each other, they heat up. The rocks may create enough heat to start a fire.

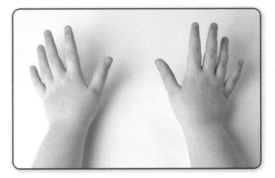

Describe the temperature of your hands.

Rub your hands together for 30 seconds. Rub them fast!

What happened to the temperature of your hands?

Before You Move On

1. What is heat?
2. Why do you use a cloth oven mitt to take a hot tray out of the oven?
3. **Predict** What would happen if you rubbed two ice cubes together?

NATIONAL GEOGRAPHIC

WIND POWER
ENERGY IN THE AIR

Many people today are talking about wind power. But wind power is not new.

Five thousand years ago, the Egyptians found that wind could help them travel. They put sails on their ships to catch the wind. The wind pushed on the sails. Boats traveled on the Nile River more quickly than ever before.

Wind gives humans energy to move ships and grind grain.

Wind power can do many things. People have used wind power to pull water out of the ground. Ancient windmills were used to crush grain to make flour.

Today windmills change the motion of wind into other kinds of energy such as electricity. The electricity made by windmills is used to power homes, office buildings, and schools. People will continue to use wind power because it is a clean, renewable source of energy.

People are turning back to wind to get energy.

These windmills on the beach of Mykonos, Greece, were once used to crush grain.

Energy is the ability to do work or cause a change. Energy comes in many forms, such as mechanical energy, sound, electricity, and heat.

Big Idea Energy comes in many forms and has the ability to do work or cause a change.

Vocabulary Review

Match the following terms with the correct definition.

A. energy

B. sound

C. heat

D. electricity

E. mechanical energy

1. The flow of energy from a warmer object to a cooler object.

2. The ability to do work or cause a change

3. An object's stored energy plus its energy of motion

4. Energy that can be heard

5. Energy that flows through wires

Big Idea Review

1. Recall What is energy?

2. Describe Use your own words to describe stored energy and energy of motion.

3. Explain Explain why sound is a form of energy.

4. Compare How do a fan and a heater use electricity?

5. Predict What happens when you rub two objects together?

6. Apply Name one way that electricity makes your life easier and explain why.

my **SCIENCE** notebook

Write About Energy

Explain Look at the two sets of strings. Which ones make a higher sound? Which ones make a lower sound? Explain why in your own words.

guitar

bass

PHYSICAL SCIENCE EXPERT: PRODUCT DESIGNER

CHAPTER 4

Product designers make all sorts of things that use energy including toys! Judy Lee uses her creativity and knowledge of energy to design new toys for kids of all ages.

Judy Lee working in her workshop.

Q: What is your job?

I'm a product designer. I have designed toys, pet products, and even food. Product design involves engineering by having to consider things such as type of energy and different materials. For example, I have to figure out the best way for a toy to use energy. Some toys need electricity to move. Other toys you move with your hand. I also have to decide which material is best for making the toy, such as plastic or wood.

Q: When did you first know you wanted to be a product designer?

When I was a kid, I loved to build things. I loved figuring out how things worked. Then I would start thinking about how I could make things better. I still think the same way today. I guess that's why I'm a product designer.

Q: What is a typical day like for you?

Each day is different. When I start a project, I talk to lots of different people. I'll take their ideas and create something, such as a toy. I'll take the toy back to them and hear what they think. Then I'll use their feedback to make the toy better. I spend a lot of time talking with people and making things.

Q: What is your favorite thing about being a designer?

I love solving problems. I love building things that use energy in different ways. It's fun to make new things.

Judy Lee builds a new toy.

TECHTREK
myNGconnect.com

Digital Library

Judy Lee uses special tools to create her toys.

139

BECOME AN EXPERT

Use Some Energy: Have Some Fun!

Toys come in all shapes and sizes. They are fun to use, and they can teach us a lot about **energy** and work.

> This pinwheel has energy of motion when you blow on it.

energy
Energy is the ability to do work or cause a change.

TECHTREK
myNGconnect.com

Student
eEdition

Digital
Library

This robot is walking. It has energy of motion.

Have you ever used a jack-in-the-box? There's a spring inside that is pushed together as you crank the handle. The tighter the spring gets, the more stored energy it has. What happens when you let the spring go? Boing! The clown pops out of the box!

The jack-in-the-box pops out when stored energy changes into energy of motion.

141

A pogo stick also uses springs. Each time you jump on the stick, the springs push together. The springs gain stored energy.

The pogo stick uses metal springs to bounce up and down.

When the springs go back to their original size, the stored energy changes to energy of motion. The energy of motion causes you to bounce in the air. The pogo stick's stored energy plus its energy of motion is its **mechanical energy** .

MECHANICAL ENERGY: POGO STICK

This pogo stick has stored energy because its springs are pushed down.

This pogo stick has changed the stored energy into the energy of motion.

mechanical energy

The **mechanical energy** of an object is its stored energy plus its energy of motion.

143

There are many toys that use **electricity** . Electric train sets are fun to set up. You can hook the tracks together in a design. When you're finished, the train will move along the track. Electricity flows through wires to the tracks. Electricity causes the train to move.

TECHTREK
myNGconnect.com

Digital Library

Electric trains are lots of fun for kids and adults.

electricity

Electricity is energy that flows through wires.

Some toys use **heat** . Heat is the flow of energy from a warmer object to a cooler object. This toy oven heats up and bakes little cakes. The heat from the warm oven flows to the cooler cake batter. The heat from the oven cooks the batter. When you take the cakes out of the oven, the heat flows from the cakes into the air. This is how they cool off. Now they are ready to eat.

Heat from a bulb inside the oven cooks the food.

heat

Heat is the flow of energy from a warmer object to a cooler object.

Have you ever played telephone? You attach two cans to a string. One person talks while the other one listens. **Sound** travels along the string. Sounds are caused by vibrations. When a person talks, the vibrations travel to your ear and make your eardrum vibrate. This is how you hear a sound.

Sounds can have high or low pitches. The girl can change the pitch of her voice while she talks into the telephone.

In a game of telephone, one girl makes a sound with her voice while the other one listens.

sound

Sound is energy that can be heard.

All of these toys use energy. Can you describe how they work? Think about what makes them move or make a sound.

You can use a mallet to make music on this xylophone. The shorter, flat pieces have higher pitches than the longer ones.

This remote-controlled car uses electricity to move.

CHAPTER 4

SHARE AND COMPARE

Turn and Talk How does your favorite toy use energy? Form a complete answer to this question together with a partner.

Read Select two pages in this section. Practice reading the pages. Then read them aloud to a partner. Talk about why the pages are interesting.

Write Write a conclusion that tells the important ideas about energy and toys. State what you think is the Big Idea of this section. Share what you wrote with a classmate. Compare your conclusions.

Draw Draw a picture of your favorite toy. Add labels to your drawing. Share your drawing with a partner. Describe how your toy uses energy to do work.

CHAPTER
5

WHAT IS LIGHT?

These raindrops are sitting on top of flower petals. The sun shines on the raindrops. You can see the reflection of other flowers in the raindrops because of light from the sun. Light is a kind of energy you can see.

TECHTREK
myNGconnect.com

Light lets you see many flowers in this single raindrop.

After reading Chapter 5, you will be able to:

- Identify and describe light energy. Explain that we need light energy to see.
  SOURCES OF LIGHT, REFLECTION, REFRACTION, ABSORPTION, SHADOWS

- Identify sources of light energy. Recognize that light sources often give off heat as well.
  SOURCES OF LIGHT

- Recognize that light travels in a straight line. **REFLECTION, REFRACTION, SHADOWS**

- Explain that light reflects off some surfaces. **REFLECTION**

- Describe the effect of refraction when light travels from one material to another.
  REFRACTION

- Classify objects as ones that absorb light and those that reflect light. Recognize that objects heat up, depending on how much light they absorb. **ABSORPTION**

- Explain why placing an object in the path of light causes a shadow. **SHADOWS**

- Science in a Snap! Explain how light reflects off some surfaces. **REFLECTION**

WHAT IS LIGHT?

These raindrops are sitting on top of flower petals. The sun shines on the raindrops. You can see the reflection of other flowers in the raindrops because of light from the sun. Light is a kind of energy you can see.

TECHTREK
myNGconnect.com

Student
eEdition

Vocabulary
Games

Digital
Library

Enrichment
Activities

Light lets you see many flowers in this single raindrop.

SCIENCE VOCABULARY

light (LĪT)

Light is a kind of energy you can see. (p. 154)

Light from the sun shines on a raindrop.

reflection (rē-FLEK-shun)

Reflection is the bouncing of light off of an object. (p. 156)

The reflection in a mirror is an example of how light bounces off a smooth, shiny object.

my
Science
Vocabulary

absorption
(ab-ZORB-shun)

light
(LĪT)

reflection
(rē-FLEK-shun)

refraction
(rē-FRAK-shun)

TECHTREK
myNGconnect.com

Vocabulary
Games

refraction (rē-FRAK-shun)

Refraction is the bending of light when it moves through one kind of matter to another. (p. 160)

When the light hits the water, the light changes direction. It bends, or refracts.

absorption (ab-ZORB-shun)

Absorption is the taking in of light by a material. (p. 162)

Absorption of light causes objects to heat up.

Where does light come from? The greatest source of light energy on Earth is the sun. But look around you. Light bulbs, fire, fireflies, and the stars are also sources of light.

Living things all around the world need energy from the sun.

Many sources of light also give off heat. When you sit in the sun, you can see the light. You can also feel its heat. You can see and feel light and heat from other light sources, too. Think about what happens when you turn on a lamp. The light bulb in the lamp gives off light. But be careful. Never touch a lit light bulb. It also gives off heat.

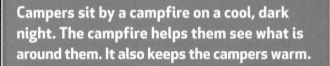

Campers sit by a campfire on a cool, dark night. The campfire helps them see what is around them. It also keeps the campers warm.

Light from the sun is natural light. Light from a light bulb is artificial light.

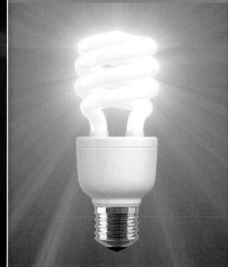

Before You Move On

1. Name some sources of light.
2. Why is the sun important?
3. Analyze How is the sun like a light bulb? How is it different?

Reflection

Have you ever seen images in a large body of water? This happens because light can bounce off of the water. A **reflection** is light bouncing off a surface. Light always moves in a straight line from the light source to that surface. Then the light bounces off the surface in another straight path.

Water is a smooth, shiny surface that reflects light. The reflection of the fall trees shows on the shiny water of the lake.

Smooth and shiny surfaces, such as mirrors, reflect most of the light that hits them. A mirror is a piece of glass with a metal coating on the back. When light hits the surface, it reflects an image that you can see.

TECHTREK
myNGconnect.com

Digital
Library

Most of the light that hits the mirror bounces back. It lets you see a clear, sharp reflection.

How People See Objects People need light to see objects. Look at this picture. The girl sees the red apple. She is able to see it because light reflects off the apple and travels in a straight line to her eyes.

The girl can see the apple because light bounces off the surface of the apple.

Science in a **Snap!** Investigating Reflection

Draw a circle on an index card. Place the index card on your desk and hold the piece of foil above the card with the shiny side facing the card.

Shine the flashlight onto the shiny side of the foil. Try to get the light from the flashlight to reflect into the circle on the card.

What did you have to do to get the light to shine in the circle?

Before You Move On

1. What happens to light when it hits a smooth, shiny surface?
2. Could you see a clear, sharp reflection in a piece of crumpled shiny wrapping paper? Why or why not?
3. **Infer** How does reflected light allow you to read?

Refraction

You've probably used a straw to drink water before. Did you ever notice that the straw can look bent from the side of the glass? As light passes from air to water, it bends, or changes direction. The straw shows that light bent at the place where it passed from air into the water. Light bending as it passes from one kind of matter to another is called **refraction** .

Refraction makes the straw look like it is in two parts.

The picture below shows a prism. A prism is a clear glass or plastic object. It often is shaped like a triangle. What happens to light that passes through a prism? The light bends as it passes through the prism. Each color that makes up the light bends a different amount. This causes light to break into the colors of the rainbow.

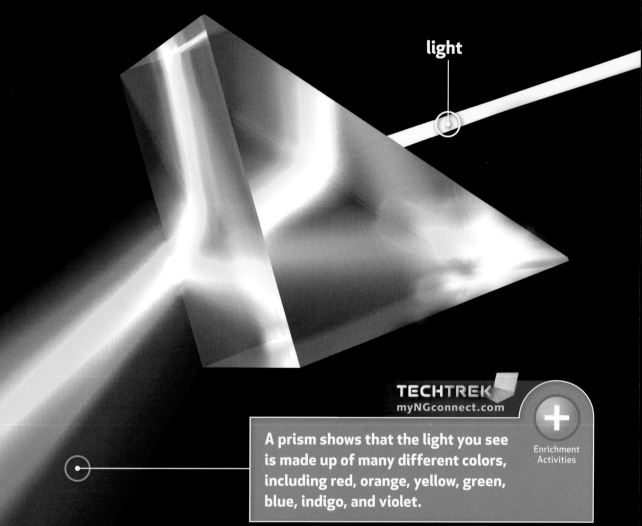

light

TECHTREK
myNGconnect.com

A prism shows that the light you see is made up of many different colors, including red, orange, yellow, green, blue, indigo, and violet.

Enrichment
Activities

Before You Move On

1. What happens to light as it passes from air to water?
2. What happens to light when it passes through a prism?
3. **Analyze** How is refraction different from reflection?

Absorption

Have you ever touched a black car on a hot day? The surface can get really hot. A black object absorbs all of the light that hits it. **Absorption** is the taking in of light by a material. The more light an object absorbs, the faster the object heats up.

The surface of the black car absorbs more light than the surfaces of the other cars.

A white object does not absorb much light. Look at the photos. If you are out in the hot sun, which cap would you wear? If you want to stay cool, you would choose the white cap. The dark cap will get hot faster than the white cap.

TECHTREK
myNGconnect.com

Digital Library

The dark-colored cap absorbs more light than the light-colored cap.

Before You Move On

1. What is absorption?
2. Would a dark blue object or light blue object absorb more light?
3. **Evaluate** You are outside in the sun in white clothes. You feel comfortable. Your friend is wearing black. He is complaining about being hot. What is happening?

Shadows

You need three things to make a shadow: a light source, an object, and an area where the light cannot reach.

You can make hand shadows by blocking the light from a flashlight. Since light always travels in a straight line, the light cannot curve around your hand. A shadow forms in the area where the light cannot reach.

TECHTREK
myNGconnect.com

Digital Library

The shadow has formed on the wall because the light could not reach it.

At noon, the sun is high in the sky. Shadows are short. Early in the morning and later in the day, the sun is lower in the sky. Shadows get longer. What can you tell about the time of the day from the shadows of the children?

The long shadows mean the sun is low in the sky. It is early morning or late in the afternoon.

Before You Move On

1. What three things do you need to make a shadow?
2. It's a sunny day. You stand on the sidewalk at noon. You stand in the same place in the early evening. Your shadow is different each time. Why?
3. **Infer** How can you make a shadow disappear?

NATIONAL GEOGRAPHIC

LIGHT POLLUTION
IN THE SKY

Lights off! See the stars in the night sky.

Lights on! Light reflects off tiny particles in the atmosphere. This causes the glow that blocks the view of the stars in the sky.

As cities grow, their lights are blocking our view of the night sky. We can't see the stars because city lights are so bright. This is called light pollution.

Light pollution harms wildlife. Some animals are active at night. For example, sea turtles lay their eggs at night. However, turtles will not crawl up onto beaches that are brightly lit. As a result, they do not lay their eggs. Some turtle populations are endangered because the turtles don't have dark beaches on which to lay their eggs.

Light pollution can be reduced by using lights that are designed to keep light from shining upward and outward. Decreasing light pollution will keep the night sky dark and will benefit many living things.

Sea turtles make nests on beaches at night and lay their eggs.

Light is energy you can see. Light can be reflected, refracted, or absorbed, depending on what object it hits.

Big Idea Light energy travels in a straight line and comes from sources that give off light and often give off heat.

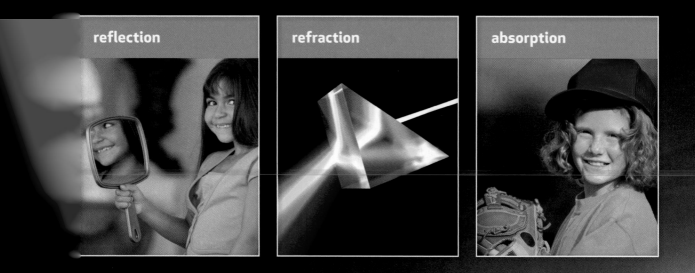

reflection

refraction

absorption

Vocabulary Review

Match the following terms with the correct definition.

A. light   **1.** The bouncing of light off of an object

B. reflection   **2.** The taking in of light by a material

C. refraction   **3.** A kind of energy you can see

D. absorption   **4.** The bending of light when it moves
                         through one kind of matter to another

Big Idea Review

1. **Cause and Effect** What happens when light hits an object?

2. **Recall** How do people see objects?

3. **Analyze** How can a mirror help you see what's behind you?

4. **Explain** Why does light cause some materials to heat up faster than others?

5. **Predict** What will happen when an object blocks the path of light?

6. **Apply** Imagine putting a pencil into a glass of water. Why might the pencil look like it bends?

Write About Refraction

Explain What is happening in this photo? How is the mist from the waterfall acting as a prism?

CHAPTER 5 PHYSICAL SCIENCE EXPERT: INVENTOR

Inventor: Dr. Patricia Bath

Imagine inventing a device that could help people to see better! Dr. Patricia Bath did just that. She discovered that laser light could help.

Dr. Bath is an ophthalmologist. An ophthalmologist is a doctor who does eye surgery and treats diseases of the eye. Dr. Bath invented the laserphaco. This machine is used to remove cataracts. A cataract is a cloudy film over part of the eye. Cataracts can cause blurred vision.

Dr. Bath invented the laserphaco to remove cataracts and help people see better.

TECHTREK
myNGconnect.com

Student
eEdition

Digital
Library

Dr. Bath explained, "When I discovered I could remove cataracts with the laser, I invented a device and called it laserphaco." This is just one of the many ways Dr. Bath has helped people to see.

Dr. Bath has done a lot of other research. She has studied how to use a laser to make surgery methods better. She also works to help prevent blindness.

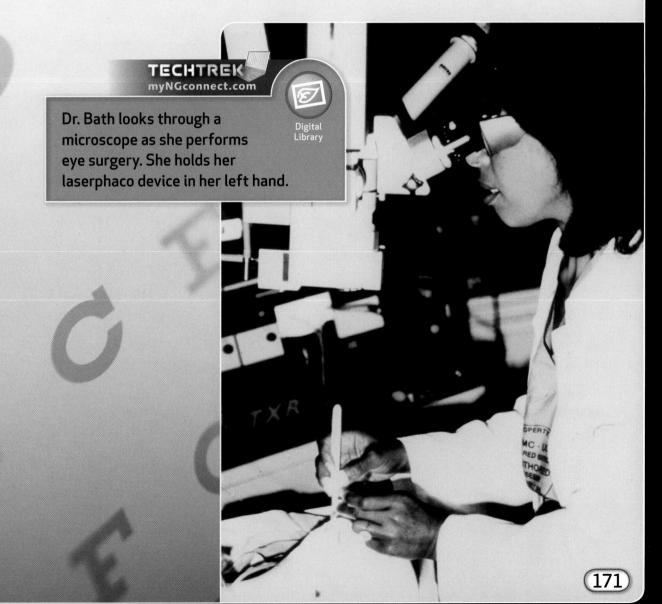

TECHTREK
myNGconnect.com

Digital
Library

Dr. Bath looks through a microscope as she performs eye surgery. She holds her laserphaco device in her left hand.

BECOME AN EXPERT

Laser: A Special Kind of Light

Laser **light** does not happen in nature. It is a special kind of light. It is made using technology.

Light from lasers can help people talk on the telephone and even use the Internet!

light
Light is a kind of energy you can see.

What makes laser light so special? It is not like regular light from a bulb. It is not like the natural light from the sun. Laser light does not have different colors of light. It is all one color. Light from the sun and light bulbs goes in all directions. Laser light goes in one direction.

TECHTREK
myNGconnect.com

Digital Library

People watch a laser light show.

The First Laser

The first laser was made in 1960. A ruby crystal was used. It was placed between two mirrors. The mirrors helped light reflect back and forth. This made a very powerful laser light. The ruby crystal showed the **reflection** of red light. The result was a ruby-colored laser light.

THE FIRST RUBY LASER

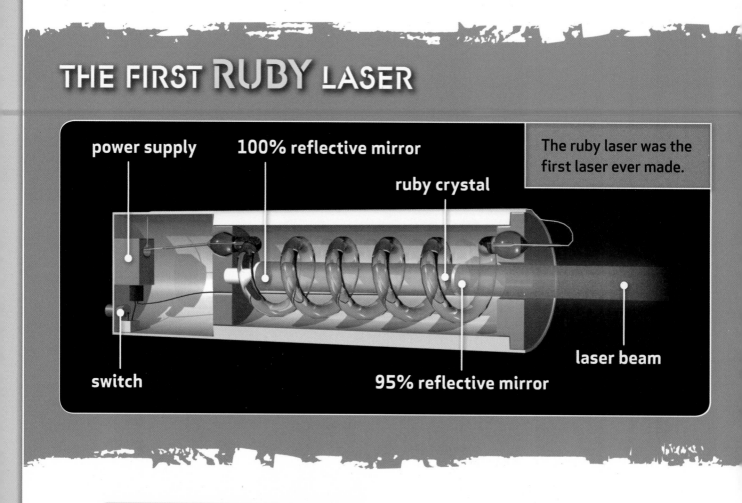

power supply

100% reflective mirror

ruby crystal

The ruby laser was the first laser ever made.

switch

95% reflective mirror

laser beam

reflection

Reflection is the bouncing of light off of an object.

What Makes Lasers Useful?

We have learned a lot about lasers since 1960. We know laser light is powerful. A laser light beam can travel great distances—even to the moon! It is also easy to control. It can be pointed at the exact spot where it is needed. These traits make lasers very useful. Lasers are used in medicine and construction. They are also used in entertainment and in outer space. Lasers are everywhere.

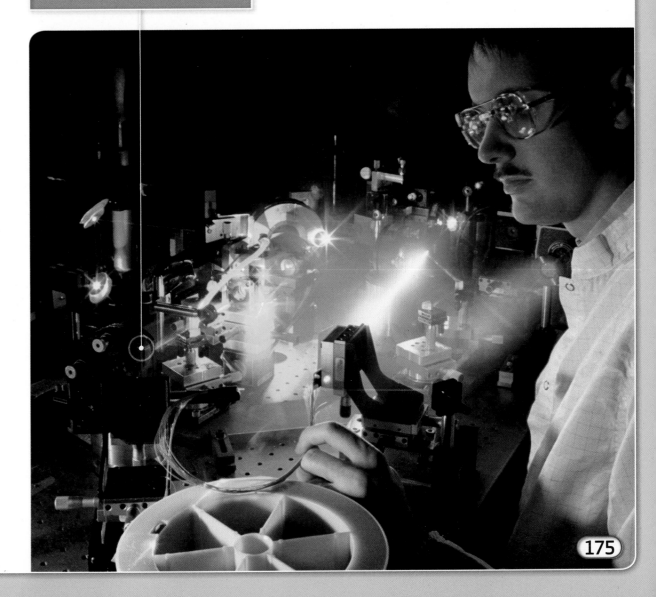

A research scientist uses a laser beam in his work.

How Eye Doctors Use Lasers

Do you know people who wear glasses? If they take off their glasses, things look out of focus or blurry. This happens because some people's eyes do not refract light correctly. **Refraction** is the bending of light. Eye doctors can use lasers to change the shape of the eye. After laser eye surgery, the eye can correctly refract the light. Then things are in focus and not blurry anymore.

refraction
Refraction is the bending of light when it moves through one kind of matter to another.

Lasers Shape Metal

A powerful laser can cut through strong metals, including steel. The metal **absorbs**, or takes in, the light energy. This makes the metal heat up until it melts. Lasers cut metal precisely. They are sometimes used to engrave small pieces of metal.

A laser cuts through steel.

absorption

Absorption is the taking in of light by a material.

Lasers All Around

Look at the many ways we use lasers in our lives.

Remember, lasers are narrow. They are also powerful. They are easy to control.

These thin cables carry laser light. The light, in turn, delivers signals such as sound, data, and images to people's computers, telephones, and televisions. Many cables can fit through the eye of this needle!

178

Lasers are used to read bar codes on items in the supermarket.

A special laser device cleans centuries of dirt from stone statues or stone ornaments on buildings without harming the stone underneath.

CHAPTER 5
SHARE AND COMPARE

Turn and Talk How would you compare laser light and sunlight? Form a complete answer to this question together with a partner.

Read Select two pages in this section. Practice reading the pages. Then read them aloud to a partner. Talk about why the pages are interesting.

my SCIENCE notebook

Write Write a conclusion that summarizes what you learned about lasers. State what you think is the Big Idea of this section. Share what you wrote with a classmate. Compare your conclusions.

my SCIENCE notebook

Draw Draw a picture of how a laser is used. Include labels. Share your drawing with a classmate. Tell how your drawings are alike and how they are different.

Glossary

A

absorption (ab-ZORB-shun)
Absorption is the taking in of light by a material. (p. 162)

C

condensation (kon-din-SĀ-shun)
Condensation is the change from a gas to a liquid. (p. 50)

E

electricity (Ē-lek-TRIS-it-ē)
Electricity is energy that flows through wires. (p. 126)

energy (EN-ur-jē)
Energy is the ability to do work or cause a change. (p. 114)

evaporation (ē-va-pōr-Ā-shun)
Evaporation is the change from a liquid to a gas. (p. 48)

F

force (FORS)
A force is a push or a pull. (p. 80)

The weightlifter uses force to pick the weights up off of the ground.

friction (FRIK-shun)
Friction is a force that acts when two surfaces rub together. (p. 84)

G

gas (GAS)
A gas is matter that spreads to fill a space. (p. 45)

gravity (GRA-vi-tē)
Earth's gravity is a force that pulls things to the center of Earth. (p. 86)

These penguins slide easily across the ice because there is not much friction.

H

heat (HĒT)
Heat is the flow of energy from a warmer object to a cooler object. (p. 130)

L

light (LĪT)
Light is a kind of energy you can see. (p. 154)

liquid (LI-kwid)
A liquid is matter that takes the shape of its container. (p. 44)

M

magnetism (MAG-na-ti-sum)
Magnetism is a force between magnets and objects magnets attract. (p. 90)

Heat flows from the warm soup to the cool spoon.

mass (MAS)
Mass is the amount of matter in an object. (p. 16)

matter (MA-ter)
Matter is anything that has mass and takes up space. (p. 10)

mechanical energy (mi-KAN-i-kul E-nur-jē)
The mechanical energy of an object is its stored energy plus its energy of motion. (p. 117)

motion (MŌ-shun)
When an object is moving, it is in motion. (p. 76)

The skier is in motion.

R

reflection (rē-FLEK-shun)
Reflection is the bouncing of light off of an object. (p. 156)

refraction (rē-FRAK-shun)
Refraction is the bending of light when it moves through one kind of matter to another. (p. 160)

S

solid (SO-lid)
A solid is matter that keeps its own shape. (p. 43)

sound (SOWND)
Sound is energy that can be heard. (p. 120)

speed (SPĒD)
Speed is the distance an object moves in a period of time. (p. 78)

states of matter (STĀTS UV MA-ter)
States of matter are the forms in which a material can exist. (p. 42)

T

texture (TEKS-chur)
Texture describes the surface of any area made up of matter. (p. 14)

V

volume (VOL-yum)
Volume is the amount of space matter takes up. (p. 18)

This man uses headphones to protect his ears from the loud sounds of the plane.

Index

Credits

Front Matter

About the Cover (bg) Chris Knapton/Photo Researchers, Inc. (t, inset) Chris Knapton/Photo Researchers, Inc. (b, inset) Alexander Gatsenko/iStockphoto.
Back Cover (bg) Chris Knapton/Photo Researchers, Inc. (tl) Courtesy of NASA/NASA Image Exchange. (tr) Photo by Bjoern Bertheau, Couresty of Synthesis Int'l. (c) Constance Adams. (bl) Constance Adams. (br) Constance Adams. **ii–iii** Paul Goldstein/Imagestate. **iv–v** BLOOMimage/Getty Images. **vi–vii** Tim Fitzharris/Minden Pictures. **vii** John William Banagan/Getty Images. **viii–ix** James Brey/iStockphoto. **ix** David Young-Wolff/PhotoEdit. **x–1** Hans Huber/Getty Images. **2** (t) Michael Rolands/iStockphoto. (c) Creatas/Jupiterimages. **2–3** vario images GmbH & Co.KG/Alamy Images. **3** (t) T.M.O. Pictures/Alamy Images. (b) Martin Barraud/OJO Images Ltd/Alamy Images. **4** (bg) Photo by Bjoern Bertheau, Couresty of Synthesis Int'l. (inset) Courtesy of NASA/NASA Image Exchange.

Chapter 1

5 Michael Rolands/iStockphoto. **6–7** Michael Rolands/iStockphoto. **8** (t) Devy Masselink/Shutterstock. (b) Rob Melnychuk/Getty Images. **9** (t) Steve Shott/Getty Images. (b) Harris Shiffman/Shutterstock. **10–11** Devy Masselink/Shutterstock. **12** Anne Rippy/Getty Images. **12–13** Jose Luis Pelaez Inc/Getty Images. **13** (inset) PhotoDisc/Getty Images. (tl) Andrew Northrup. (tr) Andrew Northrup. **14** (l) bartzuza/Shutterstock. (r) Oleg mymrin/iStockphoto. **14–15** Rob Melnychuk/Getty Images. **15** ColorBlind Images/Blend Images/Corbis. **16** Steve Shott/Getty Images. **17** (bg) Steve Taylor/Getty Images. (l) Hampton-Brown/National Geographic School Publishing. (r) Hampton-Brown/National Geographic School Publishing. **18** Carlos Criado/Shutterstock. **18–19** Harris Shiffman/Shutterstock. **19** Hampton-Brown/National Geographic School Publishing. **20–21** flashfilm/Getty Images. **21** (l) Hampton-Brown/National Geographic School Publishing. (r) Hampton-Brown/National Geographic School Publishing. **22–23** (t) Purestock/Getty Images. (b) Richard Wong/www.rwongphoto.com/Alamy Images. **23** Essdras M Suarez/Boston Globe/Landov. **24** (l) Michael Rolands/iStockphoto. (cl) Devy Masselink/Shutterstock. (cr) PhotoDisc/Getty Images. (r) bartzuza/Shutterstock. **24–25** Digital Vision/Getty Images. **25** Brand X Pictures/Jupiterimages. **26** Carol M. Grosvenor, 2009/Department of Mechanical Engineering, The University of Texas at Austin. **26–27** Pasieka/SPL/Getty Images. **27** (t) ryasick photography/Shutterstock. (b)

Carol M. Grosvenor, 2009/Department of Mechanical Engineering, The University of Texas at Austin. **28** (l) Pasieka/Photo Researchers, Inc. (c) MetaTools. (r) altiso/Shutterstock. **29** (l) Achim Prill/iStockphoto. (r) Scientifica/Visuals Unlimited. **30** (t) Antti Sompinmäki/Shutterstock. (b) motorolka/Shutterstock. **30–31** Pasieka/Photo Researchers, Inc. **31** (l) szefei/Shutterstock. (r) Ramona Heim/iStockphoto. **32** Hampton-Brown/National Geographic School Publishing. **33** (cl) Eileen Hart/iStockphoto. (cr) Pasieka/Photo Researchers, Inc. **34** E.R. Degginger/Color-Pic Inc. **35** iStockphoto. **36** Pasieka/Photo Researchers, Inc.

Chapter 2

37, 38–39 Tim Fitzharris/Minden Pictures. **40** (t) John Burcham/National Geographic Image Collection. (c) Christopher Badzioch/iStockphoto. (b) Leonard Lessin/Photo Researchers, Inc. **41** (t) PhotoDisc/Getty Images. (c) Efremova Irina/Shutterstock. (b) Creatas/Jupiterimages. **42** Michael S. Yamashita/National Geographic Image Collection. **43** (bg) Photodisc/Getty Images. (t) Christopher Badzioch/iStockphoto. (c) Jorg Greuel/Jupiterimages. (b) iStockphoto. **44** Leonard Lessin/Photo Researchers, Inc. **44–45** Brand X Pictures/Jupiterimages. **45** PhotoDisc/Getty Images. **46–47** pixel shepherd/Alamy Images. **47** (bg) John Burcham/National Geographic Image Collection. (fg) Christophe Testi/Shutterstock. **48** Efremova Irina/Shutterstock. **48–49** Lisa Stokes/Getty Images. **49** Ron Giling/Photolibrary. **50–51** Eyewire. **51** Creatas/Jupiterimages. **52** (bg) vario images GmbH & Co.KG/Alamy Images. (inset) Tom Eckerle/Getty Images. **53** (l) Andrew Northrup. (r) Andrew Northrup. **54–55** (bg) FogQuest. **55** (t) FogQuest. (c) FogQuest. (b) FogQuest. **56** (l) Nick Schlax/iStockphoto. (c) Christopher Badzioch/iStockphoto. (r) Efremova Irina/Shutterstock. **56–57** ksbell/Shutterstock. **57** Sebastian Duda/Shutterstock. **58** Patrick J. Endres/AlaskaPhotoGraphics.com. **58–59** Patrick J. Endres/AlaskaPhotoGraphics.com. **59** (t) Brice and Brice Ice Sculptures. (b) Patrick J. Endres/AlaskaPhotoGraphics.com. **60–61** Jason Lindsey/Alamy Images. **61** Arctic Images/Alamy Images. **62** Mike Goldwater/Alamy Images. **62–63** Mike Goldwater/Alamy Images. **63** (t) Mike Goldwater/Alamy Images. (bl) Stephen St. John/National Geographic Image Collection. (bc) Arctic Images/Alamy Images. (br) Mike Goldwater/Alamy Images. **64** (t) Marco Regalia Ice Hotel/Alamy Images. (b) Chad Ehlers/Alamy Images. **65** (t) E.D. Torial/Alamy Images. (b) Elisa Locci/Shutterstock. **66–67** yoel harel/Alamy Images. **68** Chad Ehlers/Alamy Images.